MICRO-LIFE

SCIENCE
EDUCATION FOR
PUBLIC
UNDERSTANDING
PROGRAM

UNIVERSITY OF CALIFORNIA AT BERKELEY
LAWRENCE HALL OF SCIENCE **LHS**

RONKONKOMA, NEW YORK

This book is part of SEPUP's middle school science course sequence:

Science and Life Issues
My Body and Me
Micro-Life
Our Genes, Our Selves
Ecology and Evolution
Using Tools and Ideas

Issues, Evidence and You
Water Usage and Safety
Materials Science
Energy
Environmental Impact

Additional SEPUP instructional materials include:

CHEM-2 (Chemicals, Health, Environment and Me): Grades 4–6

SEPUP Modules: Grades 7–12

Science and Sustainability: Course for Grades 10–12

This project was supported, in part, by the
National Science Foundation
Opinions expressed are those of the authors
and not necessarily those of the Foundation.

1 2 3 4 5 6 7 8 9   06 05 04 03 02 01
©2001 The Regents of the University of California
ISBN: 1-887725-33-4

SEPUP
Lawrence Hall of Science
University of California at Berkeley
Berkeley CA 94720-5200

e-mail: sepup@uclink4.berkeley.edu
Website: www.sepuplhs.org

Published by:

17 Colt Court
Ronkonkoma NY 11779
Website: www.lab-aids.com

A Letter to SALI Students

As you examine the activities in this book, you may wonder, "Why does this book look so different from other science books I've seen?" The reason is simple: it is a different kind of science program, and only some of what you will learn can be seen by leafing through this book!

Science and Life Issues, or *SALI,* uses several kinds of activities to teach science. For example, you will design and conduct an experiment to investigate human responses. You will explore a model of how species compete for food. And you will play the roles of scientists learning about the causes of infectious disease. A combination of experiments, readings, models, debates, role plays, and projects will help you uncover the nature of science and the relevance of science to your interests.

You will find that important scientific ideas come up again and again in different activities. You will be expected to do more than just memorize these concepts: you will be asked to explain and apply them. In particular, you will improve your decision-making skills, using evidence and weighing outcomes to decide what you think should be done about scientific issues facing society.

How do we know that this is a good way for you to learn? In general, research on science education supports it. In particular, the activities in this book were tested by hundreds of students and their teachers, and they were modified on the basis of their feedback. In a sense, this entire book is the result of an investigation: we had people test our ideas, we interpreted the results, and we revised our ideas! We believe the result will show you that learning more about science is important, enjoyable, and relevant to your life.

SALI Staff

SEPUP STAFF

Dr. Herbert D. Thier, Program Director
Dr. Barbara Nagle, Co-Director
Laura Baumgartner, Instructional Materials Developer
Asher Davison, Instructional Materials Developer
Manisha Hariani, Instructional Materials Developer
Daniel Seaver, Instructional Materials Developer
Dr. Marcelle Siegel, Instructional Materials Developer
Marlene Thier, Teacher Education and CHEM Coordinator
Dr. Peter J. Kelly, Research Associate (England)
Dr. Magda Medir, Research Associate (Spain)
Mike Reeske, Development Associate
Miriam Shein, Publications Coordinator
Sylvia Parisotto, Publications Assistant
Roberta Smith, Administrative Coordinator
Judy Greenspan, Administrative Assistant
Donna Anderson, Administrative Assistant

CONTRIBUTORS/DEVELOPERS

Barbara Nagle
Manisha Hariani
Herbert D. Thier
Asher Davison
Susan K. Boudreau
Daniel Seaver
Laura Baumgartner

TEACHER CONTRIBUTORS

Kathaleen Burke
Richard Duquin
Donna Markey

CONTENT AND SCIENTIFIC REVIEW

Jim Blankenship, Professor and Chairman of Pharmacology, School of Pharmacy, University of the Pacific, Stockton, California

Peter J. Kelly, Emeritus Professor of Education and Senior Visiting Fellow, School of Education, University of Southampton, Southampton, England

Deborah Penry, Assistant Professor, Department of Integrative Biology, University of California at Berkeley, Berkeley, California

Arthur L. Reingold, Professor, Department of Public Health Biology and Epidemiology, University of California at Berkeley, Berkeley, California

RESEARCH ASSISTANCE

Marcelle Siegel, Leif Asper

PRODUCTION

Project coordination: Miriam Shein
Production and composition: Seventeenth Street Studios
Cover concept: Maryann Ohki
Photo research and permissions: Sylvia Parisotto
Editing: WordWise

Field Test Centers

The classroom is SEPUP's laboratory for development. We are extremely appreciative of the following center directors and teachers who taught the program during the 1998–99 and 1999–2000 school years. These teachers and their students contributed significantly to improving the course.

REGIONAL CENTER, SOUTHERN CALIFORNIA
Donna Markey, *Center Director*
Kim Blumeyer, Helen Copeland, Pat McLoughlin, Donna Markey,
Philip Poniktera, Samantha Swann, Miles Vandegrift

REGIONAL CENTER, IOWA
Dr. Robert Yager and Jeanne Bancroft, *Center Directors*
Rebecca Andresen, Lore Baur, Dan Dvorak, Dan Hill, Mark Kluber, Amy Lauer,
Lisa Martin, Stephanie Phillips

REGIONAL CENTER, WESTERN NEW YORK
Dr. Robert Horvat, *Center Director*
Kathaleen Burke, Mary Casion, Dick Duquin, Eleanor Falsone, Lillian Gondree,
Jason Mayle, James Morgan, Valerie Tundo

JEFFERSON COUNTY, KENTUCKY
Pamela Boykin, *Center Director*
Charlotte Brown, Tara Endris, Sharon Kremer, Karen Niemann,
Susan Stinebruner, Joan Thieman

LIVERMORE, CALIFORNIA
Scott Vernoy, *Center Director*
Rick Boster, Ann Ewing, Kathy Gabel, Sharon Schmidt, Denia Segrest,
Bruce Wolfe

QUEENS, NEW YORK
Pam Wasserman, *Center Director*
Gina Clemente, Cheryl Dodes, Karen Horowitz, Tricia Hutter, Jean Rogers,
Mark Schmucker, Christine Wilk

TUCSON, ARIZONA
Jonathan Becker, *Center Director*
Peggy Herron, Debbie Hobbs, Carol Newhouse, Nancy Webster

INDEPENDENT
Berkeley, California: Robyn McArdle
Fresno, California: Al Brofman
Orinda, California: Sue Boudreau, Janine Orr, Karen Snelson
Tucson, Arizona: Patricia Cadigan, Kevin Finegan

Contents

UNIT C Micro-Life

30 MODELING
It's Catching! — C-4

31 PROJECT
The Range of Disease — C-8

32 INVESTIGATION
Who Infected Whom? — C-12

33 VIEW AND REFLECT
From One to Another — C-17

34 TALKING IT OVER
The Story of Leprosy — C-19

35 LABORATORY
A License to Learn — C-22

36 LABORATORY
Looking for Signs of Micro-Life — C-27

37 ROLE PLAY
The History of the Germ Theory of Disease — C-31

38 LABORATORY
Microbes, Plants, and You — C-42

39 LABORATORY
Cells Alive! — C-47

40 MODELING
A Cell Model — C-51

41 MODELING
A Cell So Small — C-55

42 READING
A Closer Look — C-58

43 LABORATORY
Microbes Under View — C-64

44 INVESTIGATION
Who's Who? — C-68

45 READING
The World of Microbes — C-70

46 INVESTIGATION
Disease Fighters — C-77

47 LABORATORY
Reducing Risk — C-82

48 INVESTIGATION
Wash Your Hands, Please! — C-85

49 ROLE PLAY
An Ounce of Prevention — C-90

50 VIEW AND REFLECT
Fighting Back — C-97

51 MODELING
The Full Course — C-99

52 TALKING IT OVER
Miracle Drugs—Or Not? — C-103

53 INVESTIGATION
Modern Outbreaks — C-107

Micro-Life

C

Unit C

Micro-Life

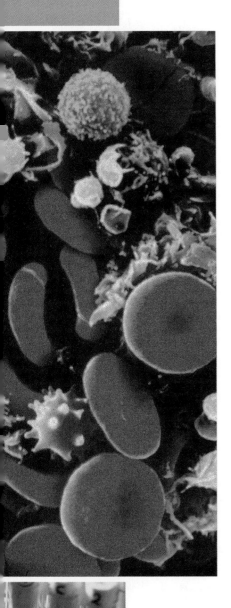

It was summer and Alex was spending two weeks at camp. In the middle of the first week, he discovered small spots all over his skin. The spots were so itchy that he went to the camp nurse.

"This rash I've got is really bothering me. Could I have some lotion?" Alex asked.

The nurse looked Alex over carefully. "I'm afraid you're going to need more than a little lotion. You've got the chicken pox!"

"What do you mean? Is that bad? I do feel a little sick, but I thought it was from the camp food," grinned Alex.

"Don't worry, you'll be fine," reassured the nurse. "But I know from the health records that there are people at camp who have never had the chicken pox nor have been vaccinated against it. So I'm going to have to isolate you in the sick room until your mom comes to take you home."

"Can my new friends visit me, at least?"

"I'm afraid not. To be safe, only the camp staff members we know have had the chicken pox will be allowed near you."

• • •

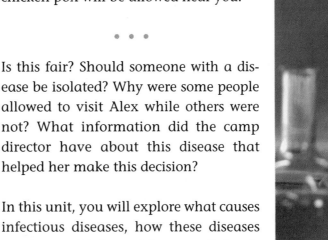

Is this fair? Should someone with a disease be isolated? Why were some people allowed to visit Alex while others were not? What information did the camp director have about this disease that helped her make this decision?

In this unit, you will explore what causes infectious diseases, how these diseases are transmitted, and how medicine is used to combat them.

30 It's Catching!

MODELING

Different diseases are caused by different factors, such as germs, heredity, or even the environment. Some diseases caused by germs are **infectious** (in-FEK-shuss), which means that they can be passed from one person to another. Many infectious diseases, such as chickenpox, are more common among children. How quickly can an infectious disease spread among a group of people? What can be done to stop more people from getting sick?

CHALLENGE

How does an infectious disease spread in a community?

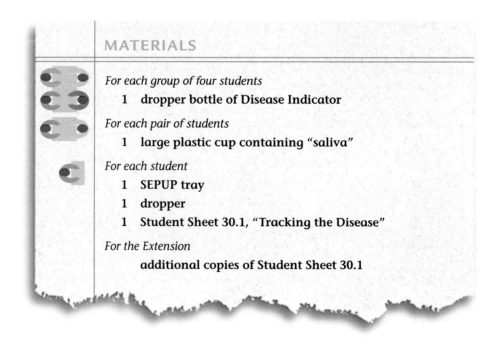

MATERIALS

For each group of four students
   1 dropper bottle of Disease Indicator

For each pair of students
   1 large plastic cup containing "saliva"

For each student
   1 SEPUP tray
   1 dropper
   1 Student Sheet 30.1, "Tracking the Disease"

For the Extension
     additional copies of Student Sheet 30.1

PROCEDURE

Part One: Planning Your Day

1. In Table 1, "My Movements" (on Student Sheet 30.1, "Tracking the Disease"), fill in the Place column by listing the one place or event that you plan to go to each day.

REMINDER

Good laboratory procedure means no accidental contamination! Be sure to follow the directions and be careful with your dropper.

2. Use your dropper to put 10 drops of "saliva" from the large plastic cup into large Cup A of your SEPUP tray.

3. Use your dropper to fill Cup B ¾-full of "saliva" from the large plastic cup.

4. After you and your partner have completed Steps 2 and 3, return the large plastic cup to your teacher.

Part Two: Meeting Other People

Your teacher will guide you through Steps 5–8.

5. On Day 1, go to the place you chose and recorded in Table 1 on Student Sheet 30.1. Then:

   a. Read the card to determine the number of people with whom you should exchange "saliva." (If no one else is at this place, you do not need to do anything.)

   b. Exchange "saliva" with people at this place by using your dropper to transfer 10 drops of solution from your Cup B into Cup B of the other student's tray while the other student transfers 10 drops of solution from his or her Cup B into Cup B of your tray. Cup B should now contain about the same amount of solution with which you started.

   c. Use your dropper to remove half of the solution from your Cup B and place it into Cup C of your own SEPUP tray.

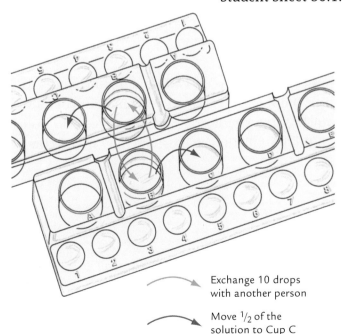

Exchange 10 drops with another person

Move ½ of the solution to Cup C

Activity 30 • It's Catching!

6. On Day 2, go to the place you chose in Table 1. Repeat Steps 5a and 5b, but this time, transfer solutions in Cup C.

7. Use your dropper to remove half of the solution from your Cup C and place it into Cup D of your own SEPUP tray.

8. On Day 3, go to the place you chose in Table 1. Repeat Steps 5a and 5b, but this time, transfer solutions in Cup D.

Part Three: Getting Tested

9. *Did you catch the disease?* Find out by testing Day 3 (Cup D) by adding 2 drops of Disease Indicator.

   If you have been infected with the disease, the solution will change color. If the solution does not change color, congratulations—you have escaped catching the disease this time! Record your results in Table 1 on Student Sheet 30.1.

10. *If you were infected with the disease, when did you get it?* Find out by testing your initial "saliva" (Cup A), Day 1 (Cup B), and Day 2 (Cup C). Record your results in Table 1 on Student Sheet 30.1.

11. Use the class data to complete Table 2, "Class Results," on Student Sheet 30.1.

12. Use the data in Table 2 to create a line graph of the number of infected people over time. Be sure to include the initial data (Day 0), to label your axes, and to title your graph.

EXTENSION

As a class, repeat the activity. Be sure to choose different places to visit or events to attend. Did the disease spread within your community in the same way? What similarities or differences do you observe? What role do the initially infected people play in affecting the spread of disease?

ANALYSIS

1. Use your graph of the class results to answer the following questions:

    a. What happened to the number of people infected with the disease over time?

    b. How does this compare to your initial prediction? Explain.

2. Think about how the infectious disease was spread from person to person in your community. If you were trying to avoid catching the disease, what could you do? Use evidence from this activity to support your answer.

3. **a.** Imagine that you are the director of the health department in the town where this disease is spreading. It is your job to help prevent people from getting sick. Explain what you would recommend to try to prevent more people from getting infected.

    b. What are the trade-offs of your recommendations?

4. What are the strengths and weaknesses of this model for the spread of infectious diseases?

5. Could you use this activity to model how diseases that are *not* infectious are spread? Explain.

31 The Range of Disease

PROJECT

You walk down the street and see a billboard that warns you about the risks of young people smoking. You read a magazine ad that tells you to drink more milk. You turn on the TV and watch a 10-second spot encouraging you to read more. These ads, which aren't selling a specific brand or service, are known as public service announcements (PSAs). PSAs provide useful and important information to the public. Many PSAs encourage children and younger adults to make choices to ensure long-term health. People who put out PSAs are responsible for making sure that the information is accurate and helpful. This means that claims should be supported by research or scientific studies. In this activity, you will make a PSA about a disease.

Disease is simply a breakdown in the structure or function of a living organism. In humans, there are many different ways that our structures (such as tissues and organs) and our functions (such as digestion) can be affected. As a result, many different diseases can affect people. Before you make your PSA, you will need to learn about a disease and decide what information is important to share with others.

Many PSAs encourage younger adults and children to make choices that ensure long-term health.

The Range of Disease • Activity 31

What type of information should be presented in a PSA on a disease?

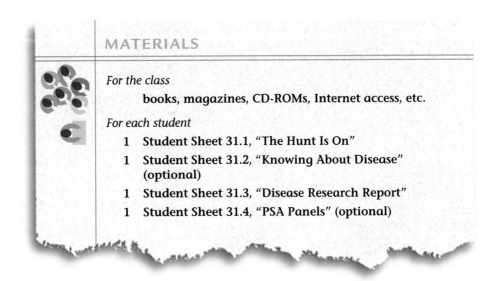

MATERIALS

For the class
   books, magazines, CD-ROMs, Internet access, etc.

For each student
   1   Student Sheet 31.1, "The Hunt Is On"
   1   Student Sheet 31.2, "Knowing About Disease" (optional)
   1   Student Sheet 31.3, "Disease Research Report"
   1   Student Sheet 31.4, "PSA Panels" (optional)

PROCEDURE

Part One: Knowledge of Disease

1. Use Student Sheet 31.1, "The Hunt Is On," to find people in your class who know someone who has had a particular disease. Have them initial the box with the name of the disease. Each person (including you) can initial only one box on your Student Sheet.

2. On Student Sheet 31.2, "Knowing About Disease," mark an "X" in the "Student" column if you know someone who has had that disease.

3. Use Student Sheet 31.2 to find out whether a parent/guardian and/or a grandparent/older adult have known someone with a particular disease.

4. As a class, total the number of students, parents/guardians, and/or grandparents/older adults who have known someone with a particular disease.

Activity 31 • The Range of Disease

Part Two: Disease Research

5. Choose a disease to research. You may want to choose from the list of diseases on Student Sheet 31.2.

6. Over the next few days or weeks, find information on this disease from books, magazines, CD-ROMs, the Internet, and/or interviews. You can also go to the SEPUP website to link to sites with more information on diseases mentioned in this activity.

7. Use the information you find to complete Student Sheet 31.3, "Disease Research Report." You should be able to describe

   a. what causes this disease

   b. symptoms of this disease

   c. how this disease is spread among humans

   d. how this disease can be prevented

   e. how this disease is medically treated

   f. two important and/or interesting facts about this disease.

8. Use your Disease Research Report to create a public service announcement (PSA) in the form of a cartoon strip. Develop a 3–6 panel cartoon strip that tells people either how to prevent getting the disease or at least one important piece of information about the disease. An example is shown below. Remember, you can use humor, but be sure your PSA is appropriate for the classroom!

ANALYSIS

Part One: Knowledge of Disease

1. For which diseases was it easy to find someone to initial your boxes on Student Sheet 31.1?

2. Would you expect to find that the same diseases are equally common in different parts of the world? Why or why not?

3. Compare the number of students, parents, and grandparents who knew someone with a particular disease. What patterns do you observe? For example, which diseases were more familiar to the grandparent generation than your generation? What do you think is the reason for this?

Part Two: Disease Research

Look at the PSAs produced by other students.

4. What can you do to prevent catching an infectious disease?

5. What types of diseases cannot be prevented? Explain.

32 Who Infected Whom?

Epidemiologists (eh-puh-dee-mee-AH-luh-jists) are scientists who trace the spread of a disease through a population. They do this to learn how the disease spreads and to find ways to help prevent its further spread. One way epidemiologists gather such information is by going to a community and comparing sick people with healthy people. Their work is complicated because not all people who are exposed to an infection get sick. Vaccinations, previous exposure to the infection, and overall health affect whether a person who is exposed to an infection will become sick. Sometimes there is no obvious connection among the sick individuals. Epidemiologists then have to resort to testing for the infection in healthy people to find the transmission path. This is because you can pass on an infection before you know you are sick. Some infectious diseases such as typhoid (TIE-foyd) and diphtheria (dip-THEER-ee-uh) can be carried (and spread) for a long time by someone who never develops symptoms of the illness. These **carriers** can be important links in the spread of a disease.

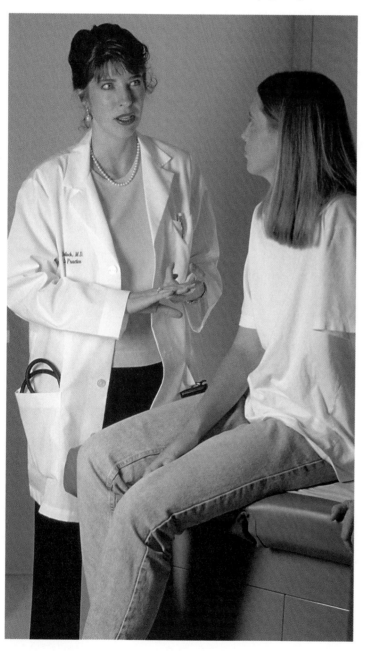

The Abingdon Chronicle

December 15

New Disease at Salk Junior High School?

Yesterday, a science teacher at Salk Junior High School was sent to the hospital with a serious illness. Lab results came back with a surprising diagnosis of a new type of infection. Why the surprise? Because this type of infection has never been observed before in Abingdon.

Ms. Shah became sick on December 3 with a high fever, sore throat, and wheezing cough. At first she thought she had the flu, but when she had difficulty breathing, she went to her doctor. Dr. Holmes of Abingdon Hospital noticed similar symptoms in another patient who is a student at Salk. When interviewed, he commented, "Ms. Shah's symptoms were very serious. She was sick for a week and in the hospital for two days. Because Ms. Shah comes into contact with many students each day, I was concerned that this disease might spread. However, there were no more cases for about 10 days. I was so relieved!"

Since then, however, several more people at the school have become sick and have had similar symptoms. Local health officials are concerned. Could this be the beginning of an outbreak? A local epidemiologist, Dr. Montagu, is working to identify the path of transmission: who infected whom? In order to determine this, she has begun to gather information. To date, she has interviewed eight individuals.

Who is (or are) the carrier(s) of the disease?

Activity 32 • Who Infected Whom?

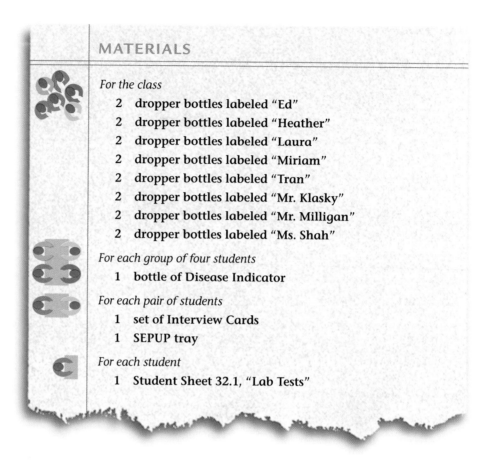

PROCEDURE

Part One: Evidence from Interviews

1. Read the information on the Interview Cards.

2. Discuss with your partner how the disease may have spread from person to person.

3. Move the cards around to develop a web showing who could have caught the disease from whom.

4. In your science notebook, draw what you think is the web of disease transmission. Be sure to include how you think these people are connected. For example:

   Ms. Shah (symptoms) $\xrightarrow{\text{science class}}$ Heather (symptoms)

5. Discuss with your partner which people may be carriers of the disease. Record what you think in your science notebook.

6. In order to test your hypothesis, you will be able to test samples of "saliva" from these people for the presence of the disease. Based on your hypothesis, record on Student Sheet 32.1, "Lab Tests," the names of four people you would most like to test.

Part Two: Collecting Lab Evidence

REMINDER

Good laboratory procedure means no accidental contamination! When using a dropper bottle, unscrew the lid but do not put the lid down on the table. Instead, use the bottle and immediately re-cap it.

7. Find the dropper bottle for one of the people you would like to test. It contains that person's "saliva." Place 3 drops of the "saliva" sample into one of the small cups in your SEPUP tray.

8. Test the sample by adding 2 drops of Disease Indicator. Make sure the dropper does not touch the "saliva." A positive test for the disease will show a pink color.

9. Record your result on Student Sheet 32.1.

10. Repeat Steps 7–9 for the rest of the people you are testing.

ANALYSIS

1. Based on your test results, draw a web showing your proposed path of disease spread. In your web, identify who is infected, the dates that he or she became sick, and whether the person is a carrier. How does this web compare to your original hypothesis?

2. a. Who was (or were) the carrier(s) of the disease?

   b. What evidence do you have to support your answer?

3. Think back to the suggestions you made to prevent the spread of disease when discussing Analysis Questions 2 and 3 of Activity 30, "It's Catching!" How does the knowledge that some diseases can be spread by carriers affect your ideas? In other words, what recommendations would you make to a community that was experiencing a disease outbreak?

4. A group of Abingdon parents have demanded that the family members and close friends of all infected individuals, including students and teachers, stay home until everyone with symptoms gets better. Explain whether you agree with their demand. Support your answer with evidence and identify the trade-offs of your decision.

Hint: To write a complete answer, first state your opinion. Provide two or more pieces of evidence that support your opinion. Then consider all sides of the issue and identify the trade-offs of your decision.

5. **Reflection:** Explain whether you would change your answer to Question 4 if the disease had more severe symptoms and a greater chance of causing death.

33 From One to Another

VIEW AND REFLECT

Humans are not the only organisms that can spread disease. Some diseases, such as the bubonic (byu-BAH-nick) plague (PLAIG) and malaria (muh-LAIR-ee-uh), are spread by vectors (VEK-terz). A **vector** is an organism (other than a person) that spreads disease-causing germs usually without getting sick itself. Rats, ticks, mosquitoes, and fleas can act as vectors for various human diseases. Ticks, for example, spread Lyme disease. That's why it's important to wear long sleeves and pants to avoid picking up ticks when hiking in some areas.

CHALLENGE

What is the role of vectors in spreading disease?

PROCEDURE

1. In order to prepare to watch the story on the video, first read Analysis Questions 1–3.

2. Watch a segment on the bubonic plague from the video, *A Science Odyssey*: "Matters of Life and Death."

ANALYSIS

1. In 1900, people did not know how the bubonic plague was spread. What did officials do to try to stop the spread of disease?

2. **a.** Draw a diagram showing how the bubonic plague is spread.

   b. Identify the vector of this disease.

3. By 1906, officials knew how the bubonic plague was spread. What did they do this time to stop the spread of disease?

4. **a.** Malaria, a disease particularly common in Africa, is caused by a tiny germ known as *Plasmodium*. When a female mosquito bites a person infected with malaria, she sucks up *Plasmodium* along with the blood. When she bites a healthy person, germs in her saliva infect that person. What is the vector in this case?

   b. Now that you know the vector of malaria, suggest two ways that the spread of malaria could be reduced or prevented.

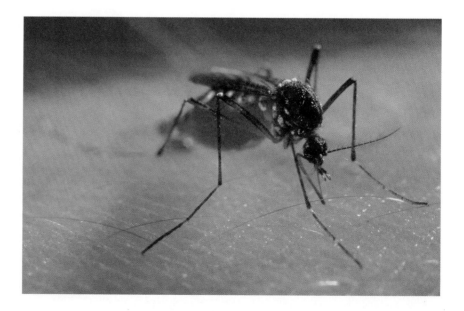

34 The Story of Leprosy

TALKING IT OVER

When the bubonic plague first occurred in San Francisco in 1900, the official response was to isolate, or quarantine (KWOR-un-teen), Chinatown. By 1906, it was clear that quarantine would not work to stop a disease spread by fleas. But during that first outbreak, people living in Chinatown were left to fend for themselves.

Leprosy (LEH-pruh-see), or Hansen's disease, is another illness that caused people to be quarantined. Because this is a long-term disease, people infected were forced to live apart from others for many years. In this activity, you will read about leprosy and make decisions about how people with leprosy and other infectious diseases should be treated.

A doctor checks the skin of a young boy receiving drug treatment to fight leprosy.

CHALLENGE

How should people with infectious diseases be treated?

PROCEDURE

1. Read the story of leprosy. As you read, think about how infectious diseases can and should be controlled.

2. Discuss Analysis Questions 1 and 2 with your group.

THE STORY OF LEPROSY

Imagine having a disease that, if people knew you had it, would cause you to be taken away from your family and forced to live somewhere else. If you had this disease, other people would be afraid to come near you or touch you. You would not be allowed to eat or sleep near uninfected people.

This is the story of leprosy. Historically, people with leprosy have been expelled from society. In the Middle Ages, people who had leprosy were considered dead. They were given a funeral service while still alive, and then forced to wander, without a home, and beg for food until they died. People have always been afraid of those with leprosy because the disease could cause serious deformities of the face, arms, and legs. Nerve damage could cause skin numbness. People sometimes lost fingers and toes to injuries that they did not feel. Damage to optic nerves could lead to blindness. In addition, there were often sores on the skin. Many of these changes were permanent and left those infected physically disabled.

In 1894, the Louisiana Leper Home was established in Carville, Louisiana. At that time, people infected with leprosy were not allowed on public transportation and were taken to the Home by boat. In fact, some individuals were placed in handcuffs and leg shackles so they would not escape. Once there, they were not allowed to make telephone calls or vote. Fear of spreading the disease meant that their outgoing mail and money were sterilized. The local soft drink company would not even accept empty bottles from the Home.

In 1873, Dr. G. A. Hansen discovered that the bacterium *Mycobacterium leprae* caused leprosy. However, no one knew how to prevent the bacteria from spreading. It was not until the late 1950s that the use of antibiotics against leprosy finally allowed people infected with the disease in the United States to choose where they wanted to live. Today, leprosy is called Hansen's disease in honor of Dr. Hansen. This new name also reflects the modern attitude toward this disease. It is now possible to treat and cure Hansen's disease with drugs. An infected person can become non-contagious after just a few days of treatment, and the spread of the disease can be controlled without long-term isolation of the victims.

The irony is that Hansen's disease does not spread easily and is very hard to catch. Only about 5% of family members living with infected people develop the disease. The exact way this disease spreads is still

The germ that causes Hansen's disease can be found in certain animals, including armadillos (shown below). Could armadillos be vectors for this disease? Scientists don't yet know the answer.

not known. Scientists believe that becoming infected requires close contact with an infected person over a long period of time. However, it has always been rare even for people caring for those with Hansen's disease to catch it. This may be because more than 90% of the population is believed to be immune; this means that these people would not become infected even after being exposed to the disease.

Although it is now rare in the United States, Hansen's disease is still a serious health problem in parts of Asia, Africa, and South America (particularly Brazil), where it usually affects the poorest people. Despite advances in its treatment, more than one new case of Hansen's disease is diagnosed worldwide each minute.

EXTENSION

Go to the SALI page of the SEPUP website to link to sites with more information about the history of leprosy in the United States.

ANALYSIS

1. How have people with Hansen's disease been treated throughout history? Provide specific examples.

2. Imagine that you meet someone who tells you that he or she has Hansen's disease. How would you respond? Support your answer with evidence from the activity.

3. Discuss what factors should determine how a person with an infectious disease should be treated.

4. Based on your understanding of infectious diseases, explain whether you think people who have an infectious disease should be quarantined. Support your answer with evidence and identify the trade-offs of your decision.

   Hint: To write a complete answer, first state your opinion. Provide two or more pieces of evidence that support your opinion. Then consider all sides of the issue and identify the trade-offs of your decision.

5. **Reflection:** In Activities 30, 32, and 33, you considered how to prevent the spread of infectious disease. Imagine that you were infected with an infectious disease. Would you volunteer to be quarantined? Explain.

35 A License to Learn

LABORATORY

Scientific discoveries often follow the development of new tools and technologies. This is certainly true in the case of infectious diseases. As you saw in Activity 33, "From One to Another," researchers Alexandre Yersin and Shibasaburo Kitasato independently used the microscope to identify the cause of the bubonic plague. Compound microscopes—microscopes that use more than one lens—were invented around 1595. These first microscopes usually magnified objects only 20–30 times their original size. But as you will learn in the next few activities, even this level of magnification was enough to discover a world of new scientific ideas.

By 1840, Italian physicist Giovanni Amici (a-MEE-chee) had invented the oil-immersion microscope which could magnify objects 6,000 times. In most middle schools, the highest level of magnification is usually about 400 times. Today, the transmission electron and scanning electron microscopes can magnify objects over 40,000 times!

What is the correct way to use a microscope?

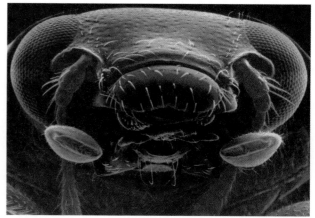

The photograph on the left shows a scanning electron microscope, which shows the surfaces of objects. In the photograph on the right, you can see the head of a ladybug as seen through this type of microscope magnified 23 times.

A License to Learn • Activity 35

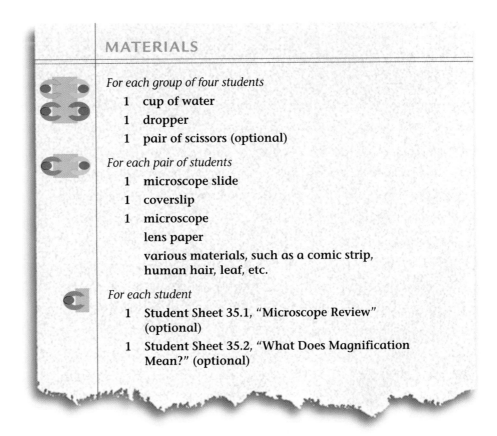

MATERIALS

For each group of four students
- 1 cup of water
- 1 dropper
- 1 pair of scissors (optional)

For each pair of students
- 1 microscope slide
- 1 coverslip
- 1 microscope
- lens paper
- various materials, such as a comic strip, human hair, leaf, etc.

For each student
- 1 Student Sheet 35.1, "Microscope Review" (optional)
- 1 Student Sheet 35.2, "What Does Magnification Mean?" (optional)

PROCEDURE

Part One: Earning a License

1. Your teacher will demonstrate the different parts of a microscope, as shown in Figure 1.

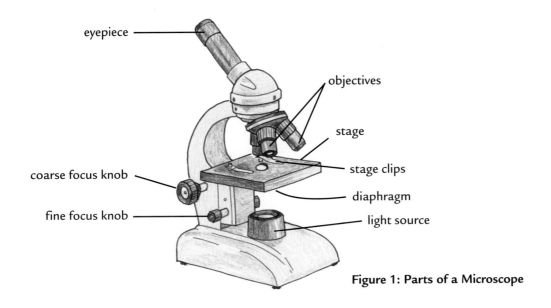

Figure 1: Parts of a Microscope

RULES FOR HANDLING A MICROSCOPE

- Always carry a microscope using two hands, as shown in the picture to the right.

- Rotate the objectives carefully. Do not allow them to touch the stage or anything placed on the stage, such as a slide. This can damage the microscope.

- When using the coarse focus knob, begin with the stage in its highest position and always focus by lowering the stage (away from the objective).

- Use only lens paper to clean the eyepiece or the objectives.

- When you have finished using a microscope, remember to turn off the microscope light and set the microscope back to low power (the shortest objective, usually 4x).

2. As a class, discuss the rules for handling a microscope.

3. Demonstrate your knowledge of the microscope, as required by your teacher.

4. Collect your microscope license! You are now ready to begin using a microscope.

Part Two: Using the Microscope

5. Clean your microscope slide and coverslip by rinsing them with water and gently wiping them dry.

6. With your partner, look at the materials list posted by your teacher and decide what you will examine under the microscope.

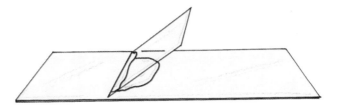

Figure 2: Placing the Coverslip

7. Place the material (or a small piece of the material) flat on the center of your microscope slide.

8. Use a dropper to place a drop of water directly onto your material. Carefully touch one edge of the coverslip to the water at an angle (as shown in Figure 2). Slowly allow the coverslip to drop into place.

9. Be sure that your microscope is set on the lowest power (shortest objective) before placing your slide onto the microscope stage. Center the slide so that the material is directly over the light opening and adjust the microscope settings as necessary.

    Hint: To check that you are focused on the material that is on the slide, move the slide slightly while you look through the eyepiece—the material that you are focused on should move at the same time you move the slide.

10. Begin by observing the sample with low power (usually the 4x objective). In your science notebook, describe how the material looks under low power compared to how it looks with your eyes. When you find an area you would like to explore at higher magnification, use the stage clips to secure the slide.

    Hint: If material on the slide is too bright to see, reduce the amount of light on the slide by slightly closing the diaphragm under the stage.

11. Without moving the slide, switch to medium power (usually the 10x objective). Adjust the microscope settings as necessary.

    Hint: If material on the slide is too dark to see, increase the amount of light on the slide by slightly opening the diaphragm under the stage.

12. When you have finished using the microscope, turn off the microscope light and set the microscope back to low power (usually the 4x objective).

ANALYSIS

1. How does the microscope change the image you see? **Hint:** Compare the material you placed on the stage with what you see through the eyepiece.

2. Describe how the material(s) that you observed looked under low as compared to medium power. What differences did you observe? How did this compare to what you saw with your eyes?

3. The microscope is one important tool used by scientists to study living things. What other tools are used by life scientists? Think about tools used by doctors and in laboratory and field research. List three tools used by life scientists and describe the kind of information they can help scientists collect.

36 Looking for Signs of Micro-Life

If someone asked you what makes you sick, you might answer that germs, bacteria, or viruses make you sick. During the early 1900s, some people thought an infectious disease like the flu could be caused by nakedness, contaminated food, irritating gases in the atmosphere, unclean clothing, open windows, closed windows, old books, dirt, dust, or supernatural causes.

What does cause infectious diseases? You can begin to answer this question with the study of **microbes** (MY-krobz), another word for creatures that are too small to be seen with the human eye. Some of these microbes cause diseases. In this activity, you will look for aquatic (water) microbes. You can see some examples of microbes in the photographs below.

What kinds of microbes can you find?

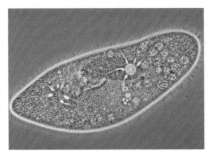

Paramecium

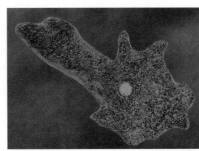

Amoeba

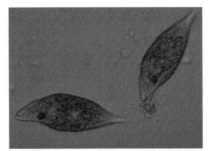

Euglena

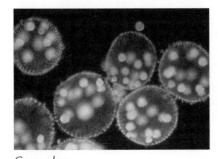

Green algae

Nematodes

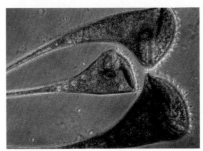

Stentor

MICROSCOPE DRAWING MADE EASY

Below is a picture taken through a microscope of the alga *Spirogyra*. The diagram to the right shows what a biologist or biological illustrator might draw and how he or she would label the drawing. Did you know that some artists draw only scientific illustrations?

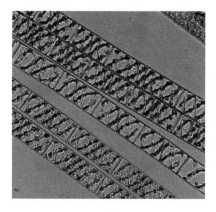
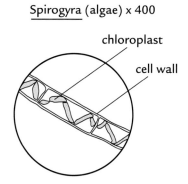

Spirogyra (algae) x 400
chloroplast
cell wall

Some tips for better drawings:

- Use a sharp pencil and have a good eraser handy.

- Try to relax your eyes when looking through the eyepiece. You can cover one eye or learn to look with both eyes open. Try not to squint.

- Look through your microscope at the same time as you do your drawing. Look through the microscope *more* than you look at your paper.

- Don't draw every small thing on your slide. Just concentrate on one or two of the most common or interesting things.

- You can draw things larger than you actually see them. This helps you show all of the details you see.

- Keep written words outside the circle.

- Use a ruler to draw the lines for your labels. Keep lines parallel—do not cross one line over another.

- Remember to record the level of magnification next to your drawing.

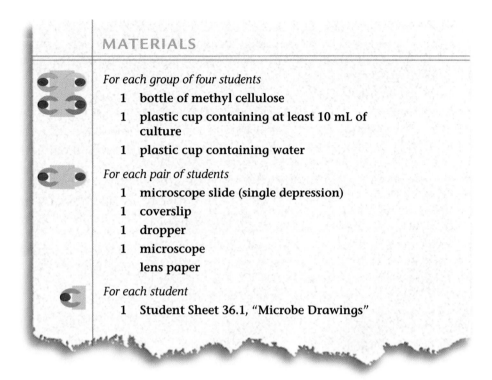

MATERIALS

For each group of four students
1 bottle of methyl cellulose
1 plastic cup containing at least 10 mL of culture
1 plastic cup containing water

For each pair of students
1 microscope slide (single depression)
1 coverslip
1 dropper
1 microscope
 lens paper

For each student
1 Student Sheet 36.1, "Microbe Drawings"

PROCEDURE

1. Clean your microscope slide and coverslip by rinsing them with water and gently wiping them dry.

2. Use the dropper in the cup containing culture to place a drop of liquid from that cup onto your slide.

3. After placing a drop of the culture liquid on the slide, add one drop of methyl cellulose directly on top of the first drop. Be careful not to add more than one drop! The methyl cellulose will slow down the movement of the microbes.

4. Carefully touch one edge of the coverslip, at an angle, to the liquid on your slide (as shown in Figure 1). Slowly allow the coverslip to drop into place.

5. Be sure that your microscope is set on the lowest power (shortest objective) before placing your slide onto the microscope. Center the slide so that the specimen is directly over the light opening and adjust the microscope settings as necessary.

Figure 1: Placing the Coverslip

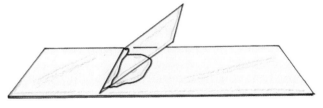

Hint: To check that you are focused on the material that is on the slide, move the slide slightly while you look through the eyepiece—the material that you are focused on should move at the same time you move the slide.

6. Begin by observing the sample on low power (usually the 4x objective). You may need to search the slide for signs of micro-life, or you may observe microbes moving through your field of view.

   Hint: If material on the slide is too light to see, reduce the amount of light on the slide by slightly closing the diaphragm under the stage.

7. Without moving the slide (which can be secured with stage clips), switch to medium power (usually 10x). Adjust the microscope settings as you look again for signs of micro-life.

8. Without moving the slide, switch to high power (usually the 40x objective). *Be careful not to smash the objective against the slide!* Adjust the microscope settings as necessary. Search the slide for signs of micro-life. Some of the microbes may be very small, so look carefully!

   Hint: If material on the slide is too dark to see, increase the amount of light on the slide: do this by slightly opening the diaphragm under the stage.

9. Review "Microscope Drawing Made Easy" on page C-28.

10. Either on medium or high power, draw at least two microbes.

11. When you have completed Step 10, turn off the microscope light and set the microscope back to low power (usually the 4x objective).

ANALYSIS

1. Is it possible that microbes exist that are smaller than those you observed? Explain how you might try to collect evidence to prove or disprove your idea.

2. As a scientist, you are asked to describe two of the microbes that you saw to someone who has never looked through a microscope. Write a short paragraph describing the microbes that you observed.

3. **Reflection:** Imagine that you are a researcher studying microbes. Would you choose to study a disease-causing microbe or one that does not cause disease? Explain.

37 The History of the Germ Theory of Disease

ROLE PLAY

Microbes, just like the ones you observed in Activity 36, "Looking for Signs of Micro-Life," were discovered in the late 1600s. But the idea that such tiny organisms could cause disease did not develop until the 1860s, less than 150 years ago. Why did it take so long to figure this out?

How did the germ theory of disease develop?

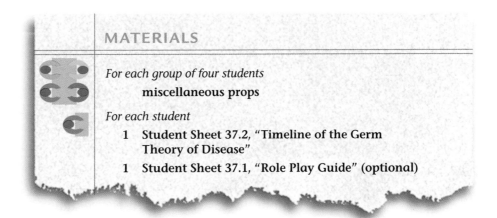

MATERIALS

For each group of four students
   miscellaneous props

For each student
1 Student Sheet 37.2, "Timeline of the Germ Theory of Disease"
1 Student Sheet 37.1, "Role Play Guide" (optional)

PROCEDURE

1. Your group will be assigned one of the sections under "Cast of Characters."

2. With your group, read the section carefully and identify the important contribution(s) to science made by your character(s).

3. Develop a skit to present these important points to your class. You can make your skit historical (for example, show how your scientist made the discovery) or modern (for example, create an ad to sell Hooke's book). Be sure to create a role for each person in your group. You can use Student Sheet 37.1, "Role Play Guide," to help you.

4. Collect any props or additional materials you require.

5. Present your skit to the class.

6. As you watch the various skits, complete the timeline on Student Sheet 37.2, "Timeline of the Germ Theory of Disease."

CAST OF CHARACTERS

Robert Hooke (1635–1703)

The late 17th century was a period of great scientific discovery. While many people offered theories without experimentation or evidence, English scientist Robert Hooke believed that good science resulted from making observations on what you could see. In his twenties, he wrote a book of his observations and drawings of the natural world called *Micrographia*, meaning "tiny drawings." It was first published in 1665. In this one book, he presented his ideas about the life cycle of mosquitoes, the origin of craters on the moon, and fossils. But Hooke is most remembered for including drawings of what he saw through a microscope.

Figure 1: Hooke's Microscope

Hooke developed his own version of the compound microscope (see Figure 1), and it was one of the best available at the time. Today, his most famous drawing from *Micrographia* is a drawing of cork—the same kind of cork that is used in corkboards and bottle stoppers. Since cork is made from the bark of the cork oak tree, it is essentially dead plant tissue. Using his microscope, Hooke looked at very thin slices of cork. He noticed what looked like little rooms (see Figure 2b). Because of this, he called these shapes **cells**, another word for *rooms*. With this simple observation, Hooke introduced an idea that would become the basis of new fields in biology—but not for almost 200 years!

Figure 2a: Cork Tree

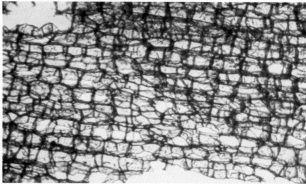

Figure 2b: Cork Cells

Anton van Leeuwenhoek (1632–1723)

Anton van Leeuwenhoek (LAY-vun-hook) was a cloth salesman in Holland and an amateur scientist. He knew how to make very simple microscopes. (Today they would be considered magnifying glasses.) But he did not become interested in studying the microscopic world until he read Hooke's *Micrographia* (see the section on Robert Hooke), which was a very popular book at the time.

Leeuwenhoek's skill at building microscopes (like the one in Figure 3) enabled him to magnify objects over 200 times. This, combined with his curiosity, led to observations almost identical to those that you made in Activity 36, "Looking for Signs of Micro-Life." In 1673, Leeuwenhoek described what he saw in a drop of water: "…wretched beasties. They stop, they stand still…and then turn themselves round…they [are] no bigger than a fine grain of sand." By examining scrapings from his teeth, he found additional evidence of these "many very little living animalcules, very prettily a-moving." Leeuwenhoek was one of the first people to observe and record microbes. He continued his observations until the end of his life.

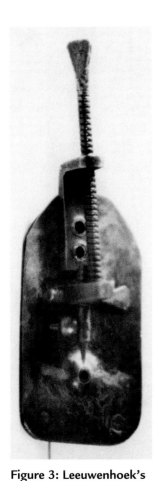

Figure 3: Leeuwenhoek's Microscope
The small hole in the board contained the magnifying lens. The material to be observed was placed on the point in front of the lens.

HOW LEEUWENHOEK DESCRIBED *SPIROGYRA*

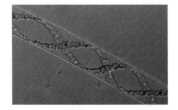

Look at the picture of the green alga *Spirogyra* shown above. Since Leeuwenhoek was not a good artist, he wrote very precise descriptions of his observations. (In addition, he hired someone to make drawings to go with his descriptions.) On September 7, 1674, he described *Spirogyra*, which can be found on lakes: "Passing just lately over this lake…and examining this water next day, I found floating…some green streaks, spirally wound serpent-wise. The whole circumference of each of these streaks was about the thickness of a hair of one's head…all consisted of very small green globules joined together: and there were very many small green globules as well."

How do Leeuwenhoek's descriptions of micro-life compare with your own?

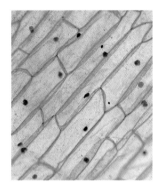

Figure 4: Plant Cells
Plants are made up of cells, like the plant cells shown in the photo above.

Matthias Jakob Schleiden (1804–1881)
Theodor Schwann (1810–1882)
Karl Theodor Ernst von Siebold (1804–1885)

Over the next 150 years, scientists continued to use the microscope to study living organisms such as plants, insects, and microbes. But by the early 1800s, most botanists—scientists who study plants—were not using microscopes. They were busy naming and describing entire plants. German biologist Matthias Schleiden (SHLY-dun) was an exception. Although he was trained as a lawyer, he left the law to become a professor of botany. Schleiden preferred to use a microscope to study plants. (Look at Figure 4 to see what Schleiden may have seen.) Based on his study, he suggested in 1838 that all plants are made of cells. This was a completely new idea: just as a house could be made up entirely of bricks, plants were made up entirely of cells!

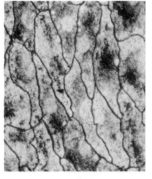

Figure 5: Animal Cells
Animals are made up of cells, like the skin cells shown in the photo above.

Schleiden knew another German biology professor, Theodor Schwann, who spent his time studying animals. Schwann was particularly interested in the digestive system. In 1839, one year after Schleiden proposed his theory, Schwann suggested that animals, and not just plants, were made up of cells. You can see animal cells in Figure 5. Because of their ideas, Schleiden and Schwann are credited with developing the **cell theory**: that all living organisms are made up of cells.

Other scientists began to build on Schleiden's and Schwann's ideas. In 1845, Karl Theodor Ernst von Siebold (SEE-bold) suggested that microbes were also made up of cells—or more specifically, one cell (see Figure 6). In fact, Siebold believed that organisms made up of many cells, like animals, were built out of single-celled microbes! While Siebold was wrong about this idea, he was right in stating that microbes were living creatures made up of the same material as animals and plants.

Figure 6: A Microbial Cell
Many microbes are made up of just one cell, like the microbe shown in the photo above.

Rudolf Carl Virchow (1821–1902)

Why is Schleiden's and Schwann's cell theory important for understanding infectious disease? Their work influenced Rudolf Carl Virchow (VIR-koh), a Polish doctor. He had been treating and studying ill patients for many years. He is famous for saying, in the 1850s, "all cells arise from cells," meaning that cells reproduce to create new cells. He was right. When you see a new plant or a baby animal, you see a **multicellular** (many-celled) creature. All living organisms begin as a single cell. Most microbes are made up of *only* a single cell,

The History of the Germ Theory of Disease • Activity 37

Did you know that each of these organisms is multicellular? Each is made up of millions of cells. You could collect evidence of this with a microscope.

as Siebold believed (see the section on Schleiden, Schwann, and Siebold). The cells of some living organisms, like people, continue to divide and grow. An adult human being is made up of about 10 trillion cells!

Virchow applied his ideas to disease. He knew that all cells grow from other cells. He thought that all diseases are caused by cells that do not work properly. He believed that diseased cells come from other usually healthy cells of the sick person. Virchow's ideas about disease were not completely correct, although they are correct for some diseases. His ideas were based on his work with leukemia (loo-KEE-mee-uh), which is a cancer of the blood. Cancer and other hereditary diseases are diseases of the cell. They *are* caused by cells that do not work properly. Infectious diseases are different, as scientists after Virchow discovered.

Ignaz Philipp Semmelweiss (1818–1865)

At the same time that Schleiden, Schwann, and Siebold were developing their ideas on cells, a Hungarian doctor working in Austria was trying to prevent young women from dying. It was the 1840s and pregnant women often died of a disease called childbed fever. Dr. Semmelweiss (sem-ul-VICE) noticed that many pregnant women were examined by doctors who had just completed an autopsy. He also observed another doctor die of childbed fever after he cut himself on the scalpel he was using to perform an autopsy. Semmelweiss concluded that childbed fever must be infectious and could be spread from something found in the dead bodies. He believed that doctors were carrying the disease from patient to patient.

Semmelweiss decided to try washing his hands between patients. As a result, fewer of his patients died. In two years, he reduced the death rate among his patients from 12% to 1%. He encouraged other doctors to use a strong chemical solution to wash their hands between patients. But because Semmelweiss could not explain why hand washing worked, many doctors refused to change their ways.

Semmelweiss tried hard to get hospitals to change their policies, but many people resisted his ideas. He eventually suffered a mental breakdown and died soon after. Within 10 years of his death, the development of the germ theory of disease would explain what he could not—that hand washing reduces the risk of infectious disease by removing germs like the ones shown in Figure 7.

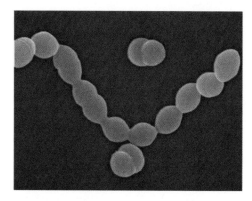

Figure 7
Today, childbed fever is called puerperal infection. It is caused by several different microbes, including Streptococcus (similar to the microbes shown to the left).

Louis Pasteur (1822–1895)

Louis Pasteur (pass-TUR), a French chemist, began studying microbes in 1864. He was working on an important business in France: the fermentation of wine and vinegar. He noticed that certain microbes could cause food and drink to spoil. Pasteur discovered that different microbes cause different kinds of spoiling, but heat can kill many of these microbes. Today, because of his work, milk is heated to 71°C for 15 seconds to kill the microbe that causes tuberculosis. Using heat to kill microbes is now known as pasteurization in Pasteur's honor. Look at Figure 8 on the next page. The word "pasteurized" on milk sold in stores tells you that the milk is safe to drink.

Figure 8: Pasteurized Milk
Pasteurization kills microbes that may be present in milk.

In 1865, Pasteur was asked to help the silk industry of France, which was having problems with silk production. Silk is produced from threads spun by a worm known as the silkworm. Pasteur observed a microbe that was infecting the silkworms and the leaves they ate. When he recommended that the worms and their food be destroyed, the silk industry was saved.

Pasteur knew that some diseases were infectious. He suggested that microbes, which he referred to as "germs," could cause infectious diseases and were easily spread by people. This idea is the basis of the **germ theory of disease**.

Robert Koch (1843–1910)

Slowly, the role of microbes in causing infectious diseases began to be accepted. But there was still more work to do. Which microbe caused which disease? In 1876, Robert Koch (KOKE), a German doctor, identified the microbe that caused anthrax (AN-thraks), an infectious disease that was killing cattle. He later went on to identify the microbes that caused tuberculosis and cholera. Amazingly, he did all of his work in the four-room apartment that he shared with his wife.

The substance inside this dish is known as agar. Many kinds of microbes can grow on agar, which provides food for the microbes.

Koch developed a way to prove that a specific microbe caused a particular disease. In the case of anthrax, he injected healthy mice with blood taken from farm animals that had died of anthrax. He injected another group of healthy mice with blood taken from healthy farm animals. All of the mice injected with the blood from the infected animals died of anthrax. None of the other group of mice developed anthrax. He then showed that he could isolate anthrax microbes only from the mice that were injected with blood from infected animals. He did not find anthrax microbes in the healthy mice. In this way, Koch was able to provide scientific evidence that the anthrax microbe caused anthrax. Figure 9 summarizes his experiment.

Koch also created new ways to grow cultures of uncontaminated microbes. In particular, he developed agar (AH-gur), a gelatin-like substance which is used to grow microbe cultures. Agar is still used today, as you will find out in Activity 47, "Reducing Risk."

Figure 9: Koch's Experiment

Healthy mice

Inject with blood from cow with anthrax

Mice die: Anthrax microbes can be isolated from blood of infected mice

Healthy mice

Inject with healthy cow blood

Mice live: No anthrax microbes in blood of healthy mice

Florence Nightingale (1820–1910)
Joseph Lister (1827–1912)
William Stewart Halsted (1852–1922)

Florence Nightingale

Ideas such as those of Pasteur and Koch were very important in the field of medicine. Florence Nightingale, an English nurse, published her ideas on disease in 1860. At the time, the idea that cleanliness was important in preventing disease was not a common one. She was one of the first to recognize the value of cleanliness and recommended it as a part of good nursing. Her efforts improved sanitary practices in military hospitals and led to fewer soldiers dying from infections due to contaminated battle injuries.

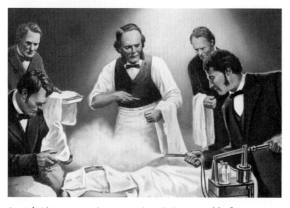

Joseph Lister supervises as antiseptic is sprayed before surgery

Scottish surgeon Joseph Lister had been concerned at the high death rates of patients following surgery. Surgery would be completed successfully, but about 45% of patients would die of infections afterward. When Lister heard about Pasteur's germ theory of disease, he came up with the idea of killing germs with chemicals. In 1867, he began using an antiseptic to clean surgical instruments. He also sprayed the air, and required hand washing and clean aprons. As a result, the death rate of patients following surgery dropped to 15%.

American surgeon William Halsted took these ideas one step further. Instead of just trying to kill the microbes once they were there, why not try to prevent them from being spread in the first place? In 1890, Halsted became one of the first surgeons to use rubber gloves during surgery. The gloves could be sterilized with heat and chemicals that were too hard on human hands. This helped reduce the presence of even more microbes and improve patient health.

By 1931, the germ theory of disease had become so accepted that ads for a disposable tissue read: "A new era in handkerchief hygiene! Use once and discard—avoiding self-infection from germ-filled handkerchiefs."

THE THEORY OF SPONTANEOUS GENERATION

Virchow stated that cells reproduce to create new cells. However, many scientists did not accept Virchow's ideas. They believed in spontaneous generation—the idea that living things grow from non-living things. For example, someone who believed in spontaneous generation might think that plants grow from soil. If you wanted to grow a plant, you would need only soil (no seeds or plant cuttings). After some time, a plant would spontaneously grow out of the soil. It took the experiments of many people to disprove the idea of spontaneous generation.

In 1668, an Italian doctor named Francesco Redi set out to show that maggots grew from eggs laid by flies. Because maggots were often found in rotting meat, many people believed that they just appeared spontaneously. To test his hypothesis, he set up several flasks containing meat. Some of the flasks were open to the air, some were sealed completely, and some were covered with gauze. As he expected, maggots appeared only in the open flasks in which the flies could reach the meat and lay their eggs.

Maggots are now known to be a juvenile stage of flies.

(continued from previous page)

In 1767, Italian priest Lazzaro Spallanzani conducted experiments to disprove spontaneous generation. He tightly sealed some bottles that contained liquid and then boiled them for more than 30 minutes. Nothing grew in the bottles. But because he had removed the air from the bottles using a vacuum, many scientists believed that Spallanzani proved only that spontaneous generation did not occur without air.

It was not until 100 years later, in 1859, that French chemist Louis Pasteur conducted a now-famous experiment that convinced most people. The French Academy of Sciences had sponsored a contest for the best experiment to either prove or disprove spontaneous generation. Pasteur's winning experiment was a variation of the method used by Spallanzani. He put a mixture of yeast, sugar, and water in several glass flasks. He then heated the necks of the flasks to bend them into the shape of an "S" (see Figure 10). Air could enter the flasks, but airborne microbes could not. Because of gravity, they would land somewhere along the neck of the flasks. Finally, he boiled the flasks to kill any microbes that might already exist in the mixtures. As Pasteur had expected, no microbes grew in the flasks. When Pasteur broke the neck of a flask and exposed it directly to air, microbes grew. Pasteur provided convincing evidence against the idea that living organisms come from non-living things.

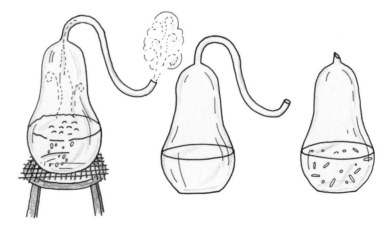

Figure 10: Pasteur's Experiment

ANALYSIS

1. Why is the germ theory of disease important in understanding infectious diseases?

2. How important was the development of the microscope in discovering the cause of infectious diseases?

3. **Reflection:** Imagine that each of the scientists in this activity wanted to hire an assistant. With which scientist would you most like to work? Why?

EXTENSION

Robert Hooke was an amazing scientist. His scientific ideas in the areas of physics, paleontology, biology, and chemistry are still relevant today. Why don't we know more about Hooke today? Some people believe it could be because the influential Sir Isaac Newton was his enemy. Find out more about Robert Hooke and his contributions to science. Begin by checking out links on the SALI page of the SEPUP website.

38 Microbes, Plants, and You

Schleiden, Schwann, and Siebold observed cells in plants, animals, and microbes. Since then, scientists have observed cells in every living organism. What do these cells look like?

What structures do different cells have in common? What structures are found only in some cells?

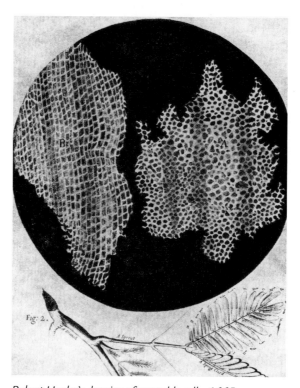

Robert Hooke's drawing of vegetable cells, 1665.

Microbes, Plants, and You • Activity 38

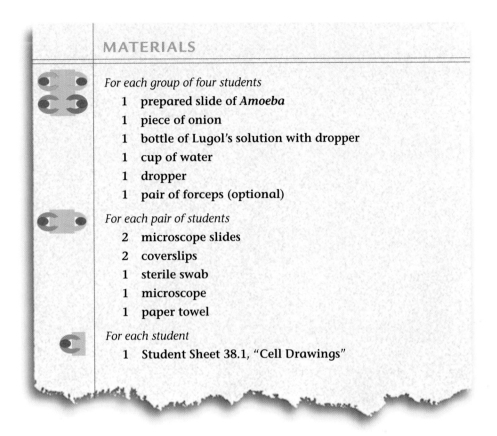

MATERIALS

For each group of four students
- 1 prepared slide of *Amoeba*
- 1 piece of onion
- 1 bottle of Lugol's solution with dropper
- 1 cup of water
- 1 dropper
- 1 pair of forceps (optional)

For each pair of students
- 2 microscope slides
- 2 coverslips
- 1 sterile swab
- 1 microscope
- 1 paper towel

For each student
- 1 Student Sheet 38.1, "Cell Drawings"

PROCEDURE

Within each group of four students, have one pair complete Parts One, Two, and Three in order. Have the other pair first complete Parts Two and Three, and then Part One.

Part One: Onion Cells

1. Use forceps or your fingernail to peel off a piece of the very thin inner layer of the onion.

2. Place 1–2 drops of water on a clean slide, then place your piece of onion in the drop of water.

3. Carefully place a coverslip over the cells. Begin by holding the coverslip at an angle over the water droplet, and then gradually lower the coverslip.

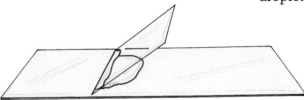

4. Be sure that your microscope is set on the lowest power (the shortest objective) before placing your slide onto the microscope stage. Center the slide so that the specimen is directly over the light opening and adjust the microscope settings as necessary.

   Hint: To check that you are focused on the material that is on the slide, move the slide slightly while you look through the eyepiece—the material that you are focused on should move at the same time as you move the slide.

5. Observe the material on the slide.

   Hint: If material on the slide is too bright to see, reduce the amount of light on the slide: do this by slightly closing the diaphragm under the stage. Move the slide so one or several of these cells are near the center of your field of view.

6. Without moving the slide (which can be secured with stage clips), switch to medium power (usually 10x). Adjust the microscope settings as necessary. Slowly focus up and down with the fine focus knob.

7. Without moving the slide, switch to high power (usually 40x). *Be careful not to smash the objective against the slide!* Adjust the microscope settings as necessary. Slowly focus up and down with the fine focus knob. You will see several layers of onion cells.

   Hint: If material on the slide is too dark to see, increase the amount of light on the slide: do this by slightly opening the diaphragm under the stage.

8. Review "Microscope Drawing Made Easy" on page C-28 in Activity 36, "Looking for Signs of Micro-Life." Select one cell to draw at high magnification. Use Student Sheet 38.1, "Cell Drawings," for your drawings. Record the level of magnification next to your drawing.

9. Return to low power and remove the slide from the microscope. Add one drop of Lugol's solution onto the slide at the edge of the coverslip. Place a small piece of paper towel on the opposite side of the coverslip (see Figure 1); this will draw the stain under the coverslip.

10. Observe the cells again at low, medium, and high power and add any new details to your drawing. Record whether you can find the edge of the cell. Record your observations of the inside of the cell.

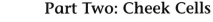

Figure 1: Staining the Slide

11. When you finish your observations, rinse the slide and coverslip and pat them dry with a paper towel.

Part Two: Cheek Cells

SAFETY

When you prepare your slide of cheek cells, each swab should be used by only one student. After spreading your cheek cells onto the slide, immediately discard your swab in the trash. Do not use it again. If you need to make another slide, use another swab.

12. Place 3 drops of water on a microscope slide.

13. Decide which person in your team of two students will volunteer to donate some cells from the inside of his or her cheek.

14. The volunteer should *gently* rub along the inside of his or her own cheek with a sterile swab. Turn the swab as you rub, making sure that each side of the swab is rubbed against the inside of your cheek. A very gentle scraping is sufficient—be sure not to cut or scratch your mouth!

15. Transfer the cheek scrapings onto the microscope slide by stirring the side of the swab in the water.

16. *Help prevent microbes from spreading!* Discard your swab in the trash as soon as you are done.

17. Carefully touch one edge of the coverslip to the water at an angle. Slowly allow the coverslip to drop into place.

18. Use Steps 4–8 as a guide to viewing and drawing your cheek cells.

19. Return to low power and remove the slide from the microscope. Add one drop of Lugol's solution onto the slide at the edge of the coverslip. Place a small piece of paper towel on the opposite side of the coverslip (see Figure 1 on the previous page); this will draw the stain under the coverslip.

20. Observe the cells again at low, medium, and high power and add any new details to your drawing. Record whether you can find the edge of the cell. Record your observations of the inside of the cell.

21. Remove the slide and follow your teacher's directions about where to put the slide for disinfecting.

Part Three: Microbe Cells

22. You and your partner should receive a microscope slide of *Amoeba*.

23. Use Steps 4–8 as a guide to viewing and drawing one *Amoeba* cell.

24. When you have completed all parts of the activity, turn off the microscope light and set the microscope back to low power.

ANALYSIS

1. Compare the three kinds of cells you have just observed.

    a. What structures do they have in common? Explain.

    b. How are the cells different? Explain.

2. Did you find evidence in this activity that the human body is made up of cells? Explain.

3. You stained the cheek and onion cells. How did the cells look before and after staining? Explain the purpose of the stain.

4. Do you think there are any small structures (organelles) inside your cheek cell other than the nucleus? What evidence do you have to support your answer?

39 Cells Alive!

Now you know that all living organisms are made up of cells. Some are made of only a single cell. Others, such as people, onions, and elephants, are made of many cells. What do all these cells do?

Large multicellular organisms such as people take in oxygen. You use the oxygen to break down nutrients. This breakdown happens in the cells in organs all over your body. When your cells break down nutrients, wastes such as carbon dioxide are produced. In the picture on the left, the swimmer's lungs have taken in oxygen and are exhaling carbon dioxide. This oxygen is used to break down sugar from food in a process called **cellular respiration**. This process provides energy your body needs and releases carbon dioxide as waste.

How do we know that all these things happen in cells? In this activity, you will investigate yeast, another type of microorganism. Yeast is a single-celled organism. Using bromthymol blue (BTB) as the indicator, you will look for evidence that yeast cells respire. You may have used BTB in Activity 17, "Gas Exchange," when you investigated your own breath. Recall that BTB can be either blue or yellow. When there is carbon dioxide in a solution, BTB is yellow. Carbon dioxide is the main waste product of cellular respiration.

Like many other living organisms, people take in oxygen from the air and produce carbon dioxide.

What do yeast cells have in common with human cells?

Activity 39 • Cells Alive!

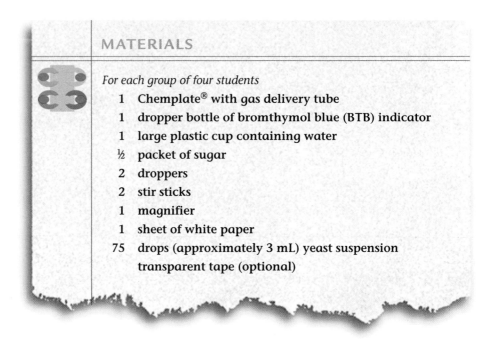

MATERIALS

For each group of four students
- 1 Chemplate® with gas delivery tube
- 1 dropper bottle of bromthymol blue (BTB) indicator
- 1 large plastic cup containing water
- ½ packet of sugar
- 2 droppers
- 2 stir sticks
- 1 magnifier
- 1 sheet of white paper
- 75 drops (approximately 3 mL) yeast suspension
- transparent tape (optional)

PROCEDURE

1. Have each person in your group of four choose one of the steps below and follow the instructions:

   a. Add 25 drops of yeast suspension to Cups 2, 3, and 4 of the Chemplate. Place this dropper near the yeast suspension. Do not use it for anything else.

   b. Add 2 drops of BTB indicator to the large oval cup of the Chemplate.

   c. Add 50 drops of water to the large oval cup of the Chemplate and stir with the BTB.

   d. Open the packet of sugar and carefully fill Cup 12 of the Chemplate with sugar. Give the rest of the sugar to another group of students.

2. Place the Chemplate on a sheet of white paper. Then use the small scoop on the end of the stir stick to add 10 scoops of sugar to the yeast suspension in Cup 3 and Cup 4.

3. Stir the sugar and yeast mixtures in Cup 3 and Cup 4.

4. Carefully observe the mixture in Cup 4 before capping it with the cup cover. Be sure that the cover fits tightly. (If not, use tape to secure it in place.)

5. Insert the tube extending off of the cup cover into the BTB solution (see Figure 1). If necessary, tape it into place.

Figure 1: Placement of the Cover and Tube

6. *Create a standard for comparison:* Place 1 drop of BTB and 25 drops of water in Cup 8.

7. Create a data table to record your initial and final observations of Cup 2, Cup 3, Cup 4, Cup 8, and the large oval cup. Be sure to list what is in each cup—for example, Cup 2 (yeast + water). Then record your initial observations of each cup, including the initial color of the solution in the large oval cup.

8. Based on your knowledge of BTB as an indicator, predict what will happen in the large oval cup. Record your prediction, making sure to explain why you think this will happen.

9. Follow your teacher's directions for observing the yeast cells through a microscope.

10. After 10–15 minutes, make observations about the liquid in the large oval cup. Then stir the solution and record its final color, making sure to compare it to the color of the standard in Cup 8.

11. Observe the mixtures in Cups 2, 3, and 4. Notice whether (and how) they have changed. Enter your observations in your data table.

ANALYSIS

 1. Compare your experimental results to your prediction. Was your prediction correct? Explain.

 2. Describe your results. Explain how your results do or do not provide evidence that yeast cells respire.

3. Think about the needs of multicellular organisms such as humans. What purpose did the sugar serve for the yeast?

4. **a.** What was the purpose of Cup 2?

   b. Imagine that you had more materials available to you. Design another control for this experiment.

 5. Based on your observations of the yeast cells under the microscope, your investigation of the gas produced by the yeast cells, and the picture of yeast cells at high magnification in Figure 2, what do yeasts have in common with humans?

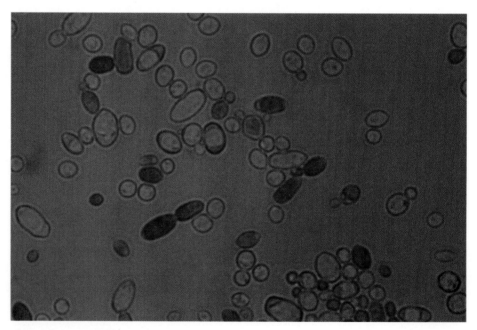

Figure 2: Yeast Cells

40 A Cell Model

People, plants, and microbes—what do they have in common? They are all made of cells. By 1846, scientists realized that cells were not hollow shapes, like balloons, but were more solid, like gelatin. You may have observed several kinds of cells. In the cheek cell and microbe cell, you could see that there was material inside the cell that can be stained. The material that fills much of the inside of cells is called **cytoplasm** (SIGH-toh-pla-zum). Every cell also has a **cell membrane** that separates it from other cells and from the environment. You can see the cytoplasm and cell membranes of the stained skin cells shown in the photo below. You were probably able to see the cell membrane of your stained cheek cells.

How does a cell membrane work? Find out by creating a simple cell model. You will use a plastic bag to model the cell membrane.

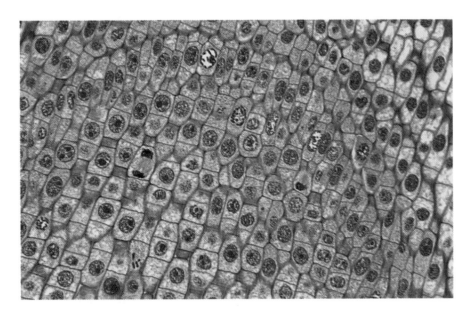

What is the function of a cell membrane?

Activity 40 • A Cell Model

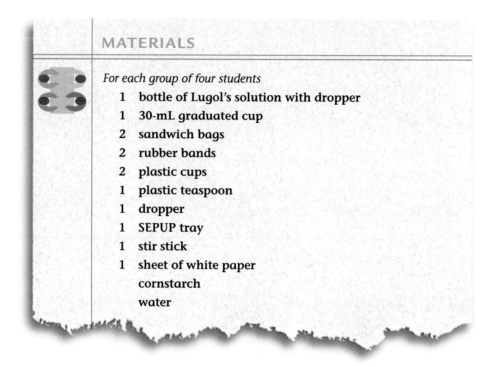

MATERIALS

For each group of four students
1   bottle of Lugol's solution with dropper
1   30-mL graduated cup
2   sandwich bags
2   rubber bands
2   plastic cups
1   plastic teaspoon
1   dropper
1   SEPUP tray
1   stir stick
1   sheet of white paper
    cornstarch
    water

PROCEDURE

1. Label the plastic cups as "Cup 1" and "Cup 2."

2. Use the graduated cup to pour 100 mL of water into each of these cups.

3. Add 7 drops of Lugol's solution to Cup 1.

4. Add 1 level teaspoon of cornstarch to Cup 2 and stir until mixed.

5. In your group of four, have one pair of students complete Step 5a, while the other pair completes Step 5b:

    a. Use the graduated cup to pour 30 mL of water into a sandwich bag. Then add 7 drops of Lugol's solution to the water in the bag.

    b. Use the graduated cup to mix 30 mL of water with one teaspoon of cornstarch. Stir and then carefully pour the mixture into a sandwich bag. Be careful to avoid getting cornstarch on the outside of the bag. If there is cornstarch on the outside of the bag, rinse the bag under cold water.

6. Use rubber bands to seal the bags.

A Cell Model • Activity 40

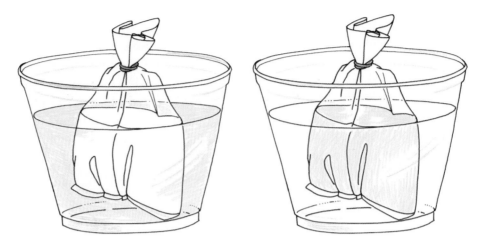

Figure 1: Initial colors of the mixtures.

7. Place the bag containing cornstarch into Cup 1 and the bag containing Lugol's solution into Cup 2, as shown in Figure 1. Then place the cups on the sheet of white paper and leave them there for 10–15 minutes. As you wait, complete Steps 8–10.

8. Create a data table to record the initial and final color of each solution, both inside and outside of the model cells in both cups. Be sure to record your initial observations.

9. Complete Steps 9a–c to find out how Lugol's solution reacts with starch.

   a. Place 5 drops of water into Cup 1 and 5 drops of water into Cup 2 of the SEPUP tray.

   b. Use the stir stick to add 1 scoop of cornstarch into Cup 2 and stir.

   c. Add 1 drop of Lugol's solution to each cup.

   d. In your science notebook, record the color of Lugol's solution when starch is present.

10. Complete Analysis Question 1.

11. After 10–15 minutes (or longer), lift the bags out of the cups and look carefully at all of the solutions. Record any changes that have occurred either in the bags or in the cups.

Activity 40 • A Cell Model

EXTENSION

Model a cell by using a real membrane from an egg. An egg can be "de-shelled" by soaking it in vinegar, leaving the rest of the egg intact. Be careful, the "de-shelled" egg is fragile. You can then place the de-shelled egg in different liquids, such as water, food coloring, paint, or corn syrup. Leave the egg in a solution for several days to find out if particles pass through the membrane. Collect data on these changes by measuring the mass of the egg before and after its soak.

ANALYSIS

1. **a.** Draw a diagram of the cell model used in this activity.

   b. Label the part of the cell model that represents the cell membrane and the part that represents cytoplasm.

   c. Label the part of the model that represents the environment outside the cell.

2. Review your results. Describe which part(s) of the lab set-up showed a reaction between Lugol's solution and starch.

3. Summarize your results by answering the following questions:

   a. Which particles—starch or Lugol's—were able to cross the model cell membrane? Explain how the experimental evidence supports your answer.

   b. Which particles—starch or Lugol's—were *unable* to cross the model cell membrane? Explain how the experimental evidence supports your answer.

4. Based on your cell model, what is the function of the cell membrane?

5. Think about the fact that cells are alive. Why is it important for particles to be able to pass through the cell membrane?

41 A Cell So Small

MODELING

Some organisms, like bacteria, consist of only one cell. Other organisms consist of several to many cells. An adult human being is made up of approximately 10 trillion cells. One drop of human blood, has about 500 *million* cells!

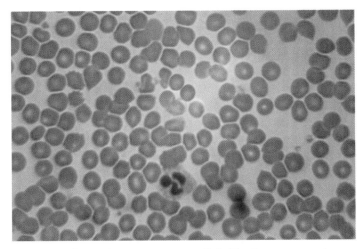

Red blood cells are the most numerous cells in blood.

Why do some cells need to be so small? Why aren't multicellular organisms like people made up of just one huge cell instead? Find out by modeling large and small cells.

CHALLENGE

Why are cells so small?

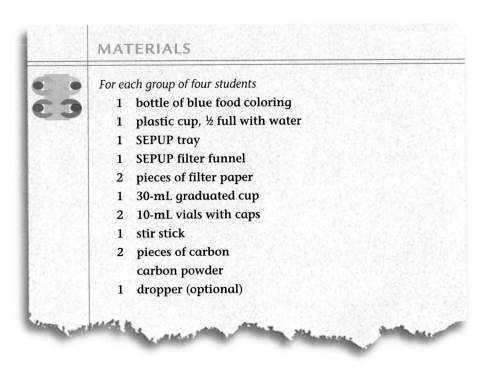

MATERIALS
For each group of four students
1 bottle of blue food coloring
1 plastic cup, ½ full with water
1 SEPUP tray
1 SEPUP filter funnel
2 pieces of filter paper
1 30-mL graduated cup
2 10-mL vials with caps
1 stir stick
2 pieces of carbon carbon powder
1 dropper (optional)

Activity 41 • A Cell So Small

REMINDER

Be careful when handling carbon. It is messy and can ruin your clothes. Never place any carbon directly onto a counter; use a piece of paper or a paper towel. Be sure to carefully clean up any spills.

PROCEDURE

1. Fold 2 pieces of filter paper into cones as shown in Figure 1: first fold each paper in half and then in half again. Open each filter paper into a cone (pull one piece to one side and push the rest to the other side).

Figure 1: Folding Filter Paper Into a Cone

2. Place the plastic SEPUP filter funnels over large Cups C and D of your SEPUP tray, as shown in Figure 2. Then place a filter paper cone into each of the funnels.

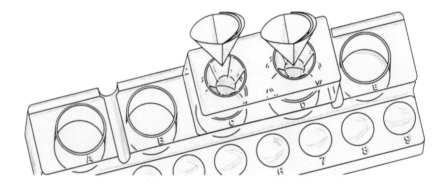

Figure 2: Setting Up the Filter

3. Dye your cup of water blue by adding 2 drops of blue food coloring. Stir.

4. *Model large cells:* Place 2 pieces of carbon into one of the 10-mL vials.

5. *Model small cells:* Using the scoop on a stir stick, your teacher will measure out the same volume of carbon powder into the other vial. You should now have the same amount of carbon in each of the two vials.

6. *Model how well the cells can take up oxygen or nutrients they need to live:* Use your 30-mL cup to add 7.5 mL of dyed water to each vial. Then cap the vials and shake each vial ten times.

7. Open the vial containing the carbon pieces. Pour the mixture through the filter paper over Cup C.

8. Open the vial containing the carbon powder. Pour the mixture through the filter paper over Cup D.

9. Observe and record the color of the water in each large cup of your SEPUP tray.

10. Clean up as directed by your teacher.

ANALYSIS

1. In this model, what did each of the following represent:
    a. carbon powder
    b. carbon pieces
    c. blue dye

2. What happened to the blue dye in each vial? Explain.

3. According to the model, which cells—large or small—are most efficient at taking up oxygen and nutrients from the environment? Explain.

4. What is one reason multicellular organisms, such as people, are made up of many small cells instead of one large cell?

42 A Closer Look

You learned that Schleiden and Schwann discovered that all living organisms are made up of one or more cells. This includes plants, animals, and many microbes. The microbes that cause infectious disease are often organisms made up of just one cell, as Siebold discovered. Most organisms you can see without a microscope are made of many cells.

The cells shown here are from human skin. What do you see inside the cells? What exactly are you looking at when you use a microscope to look at a cell?

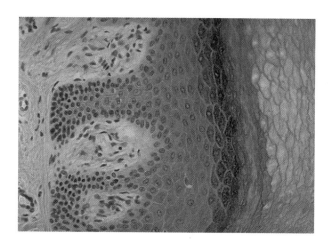
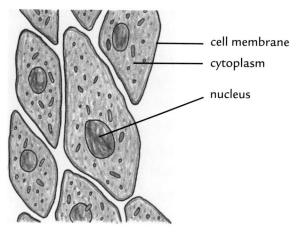

What are some of the parts of a cell? What do they do?

READING

A Typical Cell

As scientists continued to study cells, they noticed structures inside the cells. From observing many kinds of cells in thousands of organisms, biologists discovered that some structures are found in all or nearly all cells. These structures are so common that they are usually included in models or drawings of a "typical cell."

The most common structure of cells is the cell membrane (see Figure 1). This membrane separates the cell from its environment. Every kind of cell, whether a bacterium, or a cell from an elephant or from a giant oak tree, has a cell membrane. As you learned in Activity 40, "A Cell Model," the cell membrane acts as a barrier to control what enters or leaves the cell. Somehow, everything that enters or leaves the cell must cross this membrane.

The material enclosed by the cell membrane is called the cytoplasm, which means the "cell material." In the cytoplasm the cell breaks down nutrients from food and builds the new substances it needs to grow and to carry out its other functions.

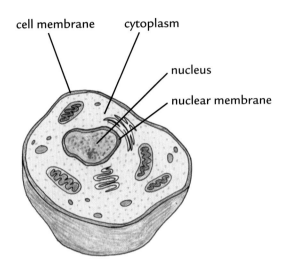

Figure 1: Inside a Cell

STOPPING TO THINK 1

a. How did scientists discover the common structure of cells?

b. What are some of the common structures of a cell?

In 1831, Robert Brown identified a small dark center within many cells. He called this center the **nucleus**. You were probably able to observe the nucleus in onion, *Amoeba*, and human cells. Most organisms—except for bacteria—have a cell nucleus. The nucleus is a small compartment within the cell. It is separated from the rest of the cell by a nuclear membrane. The nucleus contains the genetic information of the cell and directs the cell's activities, including growth and reproduction.

STOPPING TO THINK 2

a. Why is the nucleus an important part of most cells?

b. What type of organism does not contain a nucleus?

Most cells have other tiny structures that help them do many jobs. These structures are called **organelles**, or "little organs." They are often surrounded by their own special membranes. Some of the organelles can just barely be seen with a light microscope. Some of the jobs performed by these organelles include obtaining and storing energy, helping cells move and divide, and making substances that are either used in the cell or transported to other parts of the body.

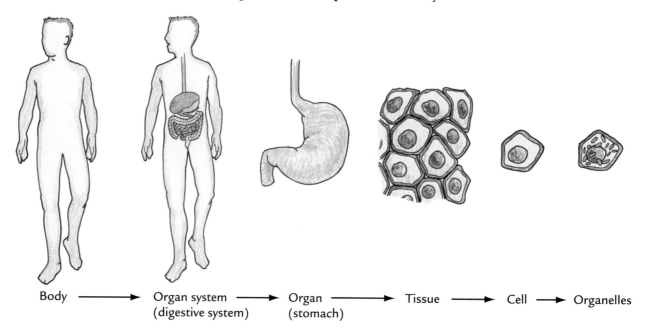

Body → Organ system (digestive system) → Organ (stomach) → Tissue → Cell → Organelles

What Can You Learn From Studying Cells?

Information about cells can be used for practical purposes, such as treating different kinds of diseases. Here are just two examples:

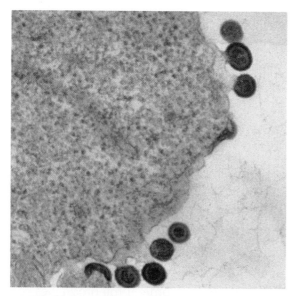

A cell biologist used an electron microscope to take this photo of HIV (the virus that causes AIDS) on the surface of a human white blood cell.

- AIDS is a disease of one group of cells within the human body. Investigating how these cells work normally and what goes wrong in the cells of a person with AIDS helps scientists understand the disease and develop treatments.

- Human blood contains red blood cells. Each red blood cell lives in the human bloodstream for only about 120 days. New red blood cells are constantly being formed as old ones die. Researchers have used information about how new red blood cells develop to prepare a drug that causes more red blood cells to form. This drug is given to patients who require certain kinds of surgery or to people with illnesses that reduce the number of red cells in the blood. The drug helps patients build up red cells and require fewer blood transfusions.

CELL BIOLOGY

Cell biology is the special branch of biology that studies cells and how they work. Cell biologists are fascinated by the variety of cells and the fantastic structures inside them. Some cell biologists study one type of cell, such as muscle cells. Others focus on special parts of cells, such as cell membranes, to understand how they work. Some of the questions that cell biologists try to answer are:

- How do the different cells in an organism work?
- What does each part of the cell do, and how?
- How do the different cells in an organism communicate and control their activities so that things happen in the right place and at the right time?
- How can one fertilized human egg cell grow into a complex adult with many kinds of cells?
- How does a cell know when to divide?

Scientists have found some exciting partial answers to these questions, but there is a great deal left to learn about cells.

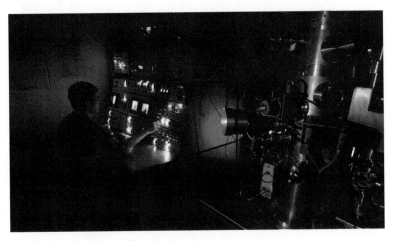

A cell biologist using an electron microscope

Activity 42 • A Closer Look

For links to photographs and more information about cells, go to the SALI page of the SEPUP website.

ANALYSIS

1. Observe the pictures of cells in Figure 2, "Animal Cells." Cells 1, 2, and 4 were taken with a scanning electron microscope which shows the surface (and not the inside) of the cell. This type of microscope magnifies the cells much more than the microscopes you use in class. You can see that the cells have quite different

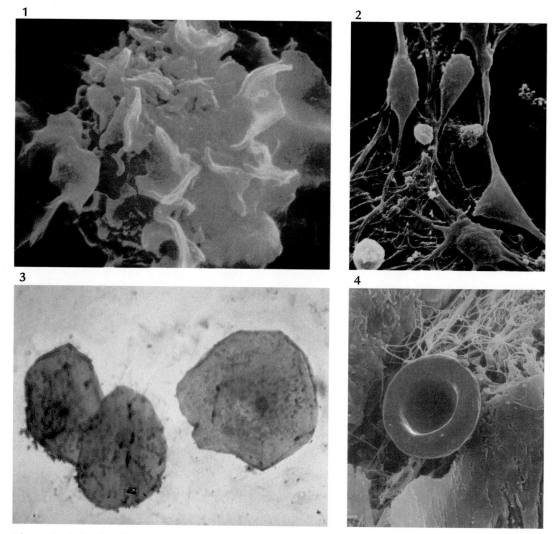

Figure 2: Animal Cells

shapes: some are rounded, while others are elongated, flat, or ruffled. These shapes depend on the cells' functions in the body. Try to match each cell with one of the following descriptions.

   a. These cells have long branching parts that send signals to distant parts of the body.

   b. These flat cells form an even covering on the surface of areas like the inside of the mouth.

   c. These round human cells are unusual because they do not have a nucleus. They are full of a protein that carries oxygen to all parts of the body.

   d. These cells are able to crawl around the body to attack bacteria and other foreign material. Ruffles on the cell membrane lead the way as the cells move.

2. Based on its description, which of the four cells described in Question 1 is a nerve cell? Which is a red blood cell? Which is a white blood cell? Which is a skin cell? Explain how you were able to match the type of cell with its function.

3. Give one example of how the study of cells helps treat diseases.

4. Explain why membranes are so important to cells.

5. Look back at your drawings from Activity 36, "Looking for Signs of Micro-Life." Did you observe any structures within the microbes that you drew? What do you think these structures are?

6. **Reflection:** Which of the questions studied by cell biologists is most interesting to you? Why?

43 Microbes Under View

Study the photographs on this page. You can see microbes of different shapes, sizes, and structures. Microbes are organized into different groups based partly on differences in their cell structure. In this activity, you will look at two different groups of microbes to see what kinds of differences you can find. You will observe stained slides of **protists** (PRO-tists) and **bacteria** (bak-TEER-ee-uh).

You first saw microbes when you looked at water samples in Activity 36, "Looking for Signs of Micro-Life." Do you recognize any of the same creatures in these photographs?

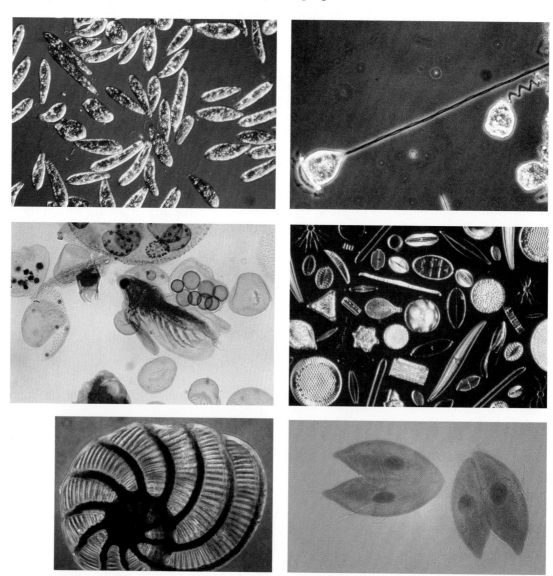

Microbes display a fascinating variety of shapes, sizes, and structures.

Microbes Under View • Activity 43

CHALLENGE

What are some of the differences among the cells of two groups of microbes?

MATERIALS

For the class
- 16 microscopes
- 4 slides of mixed bacteria
- 4 slides of *Trypanosoma* (protist)
- 4 slides of *Amoeba* (protist)
- 4 slides of *Paramecium* (protist)

For each student
- 1 Student Sheet 43.1, "Microbe Observations"

PROCEDURE

1. You and your partner should receive a microscope slide of a one-celled microbe.

2. Be sure that your microscope is set on the lowest power (shortest objective, usually 4x) before placing your slide onto the microscope stage. Center the slide so that the specimen is directly over the light.

3. Begin by observing the slide on low power. You may need to search the slide for the organisms. Be sure that an organism is in the center of the field of view (you may need to move the slide slightly) and completely in focus before going on to Step 4.

   Hint: To check that you are focused on the material that is on the slide, move the slide slightly while you look through the eyepiece—the material that you are focused on should move at the same time as you move the slide.

   Hint: On prepared slides, organisms are usually stained with dyes to make them easier to see: look for blue, purple, green, or pink organisms.

4. Without moving the slide (which can be secured with stage clips), switch to medium power (usually 10x). Adjust the microscope settings as necessary. Observe the organism.

5. Without moving the slide, switch to high power (usually 40x). *Be careful not to smash the objective against the slide!* Adjust the microscope settings as necessary.

   Hint: If material on the slide is too dark to see, increase the amount of light on the slide: do this by slightly opening the diaphragm under the stage.

6. Turn the fine focus knob up and down just a little to reveal details of the microbe at different levels of the slide.

7. Review "Microscope Drawing Made Easy" on page C-28 of Activity 36, "Looking for Signs of Micro-Life." Then draw your organism (on high power) on Student Sheet 43.1, "Microbe Observations." Be sure to record the level of magnification you are using. Make your drawing large enough to fill up most of the space on the paper. Include on your drawing details inside the cell and along the edge of the cell membrane.

8. Switch slides with another pair of students and repeat Steps 2–7.

9. Repeat Step 8 until you have seen all four microbe slides.

10. When you have completed your observations, turn off the microscope light and set the microscope back to low power.

11. Work with your group to discuss Analysis Questions 1 and 2 before the class discussion.

ANALYSIS

1. When you compare the different protists, what differences do you observe?

2. When you compare the different bacteria, what differences do you observe?

3. When you compare all of the different microbes, what similarities and differences do you observe?

4. Look at the drawings of micro-life you made for Activity 36, "Looking for Signs of Micro-Life." Could any of the organisms you saw have been protists or bacteria? Support your answer with evidence from this activity.

5. In your science notebook, create a larger version of the diagram shown below (known as a Venn diagram). Record unique features of cells of each group of organisms in the appropriate space (either "protists," "bacteria," or "human" cells). Record common features between groups in the space that overlaps. **Hint:** Think about what you have learned about cells in the last few activities. Look again at your notes from this activity.

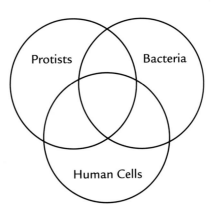

44 Who's Who?

INVESTIGATION

As you will learn in the next few activities, diseases caused by different microbes are prevented and treated differently. That's why Leeuwenhoek's discovery of microbes and Pasteur's germ theory of disease were essential. Find out more by using more evidence to classify microbes.

How are these microbes classified?

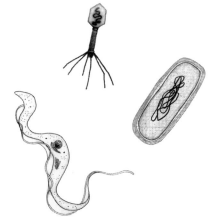

MATERIALS

For the class
   8   sets of 6 Classification Cards

For each group of four students
   1   set of 9 Micro-Life Cards

PROCEDURE

1. Spread your Micro-Life Cards out on a table. Each card shows a high magnification of the outside of the microbe and a drawing of a high magnification view of the inside.

2. Examine each card carefully, noting similarities and differences.

3. With your group members, classify the microbes into groups. Work together to agree on a classification system:

   • Listen to and consider the explanations and ideas of other members of your team.

   • If you disagree with other members of your team about how to classify a microbe, explain why you disagree.

4. In your science notebook, list the groups that you created and the common features of each group. Be sure to record which microbes belong to which group.

5. Leave your cards sorted into groups and your notebook open on your work surface. When all teams are finished you will look at what others have done.

6. View the work of other student teams. As you look at their classification systems, observe the similarities and differences between their systems and your own. Discuss your observations with your team members.

7. You will receive 6 Classification Cards from your teacher. Each card represents a group of creatures. Based on the information described on the Classification Cards, place each Micro-Life Card under one of the Classification Card categories. In your science notebook, record any changes you want to make to your original grouping of your Micro-Life Cards.

8. As a class, discuss the classification of the Micro-Life Cards. In your science notebook, record the common features of each major group.

ANALYSIS

1. How could knowing the structure and classification of disease-causing microbes help scientists fight a disease?

2. How did your system of classification compare to the Classification Cards?

3. Look back at the generalized animal cell in Figure 1 in Activity 42, "A Closer Look," on page C-59. Explain how this drawing of a cell is similar to or different from the structure of each of the following groups of microbes:

   a. protists

   b. bacteria

   c. viruses

45 The World of Microbes

Have you had a cold, flu, or other infectious disease recently? Do you know what caused your illness? Microbes cause most infectious diseases. Microbes include the protists, bacteria, and viruses that you classified in Activity 44, "Who's Who?" They also include some fungi, such as yeast and the fungi that cause athlete's foot.

By now you know that *germ* is simply another word for a microbe that causes disease. But you may have also heard the word *microorganism* used. Why, then, do we keep referring to microbes? To find out, you need to know a little more about the differences among the microbes you've studied so far (protists, bacteria, and viruses).

How do microbes fit into the classification of organisms?

READING

Classifying Organisms

Until recently, scientists classified organisms into five groups, called Kingdoms, as shown in Figure 1. New evidence has led to several alternatives to the five-kingdom system. Classification is a way to make sense of a lot of information. As the information changes, new classification systems evolve. For example, scientists have learned that bacteria can be divided into two very different groups, called Bacteria and Archaea. Still, it is useful for you to think about five different groups of organisms: animals, plants, fungi, protists, and bacteria.

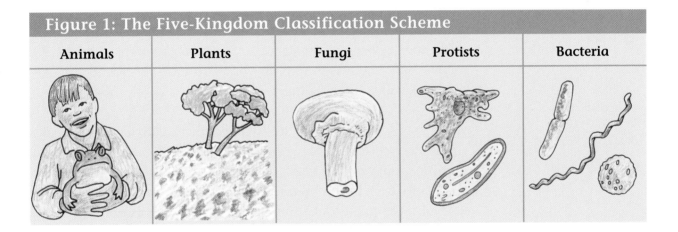

Figure 1: The Five-Kingdom Classification Scheme

| Animals | Plants | Fungi | Protists | Bacteria |

You are most familiar with animals and plants. They make up two kingdoms. A third kingdom is made up of fungi. The fungi include yeasts (like the one you used in Activity 39, "Cells Alive!"), molds, and mushrooms. Protists and bacteria, like the ones you observed in Activity 43, "Microbes Under View," belong to two more groups of organisms. Notice that viruses are not included in the figure because they are not considered to be living organisms.

STOPPING TO THINK 1

Think about all of the slides you have observed. Have you observed cells of organisms from every kingdom? List all the cells you have observed from organisms in each kingdom.

Protists

Protists are single-celled microbes that have a nucleus. While some protists cause illness, many others are harmless. The *Trypanosoma* that you observed in Activity 43 is closely related to another type of *Trypanosoma* that causes sleeping sickness in people. Species of *Paramecium* are often harmless, living in fresh and salt water, where they feed on bacteria, algae, and other protists. Many types of *Amoeba* are harmless, while others cause illnesses of the digestive system.

Bacteria

Bacteria are single-celled microbes that do not have a nucleus. Bacteria are also the most common microbes and can be found everywhere—in snow, deserts, lakes, the ocean, and the human body. As you may recall, bacteria are extremely tiny; a thousand bacteria could fit in a cluster on the dot of an "i." There are more bacterial cells in your digestive system and on your skin than the number of cells that make up your entire body!

While some bacteria, such as *Mycobacterium tuberculosis*, cause diseases, other species of bacteria are helpful. In fact, without bacteria, nothing would ever decompose; the world would be full of dead organisms, from the tiniest microbes to large plants and animals! Bacteria also are important in the preparation of foods and beverages. You may have noticed a statement on some yogurt containers: "contains live and active yogurt cultures." That's because yogurt is produced by the fermentation of milk by bacteria! Figure 2 shows the shapes and some information about different kinds of bacteria.

STOPPING TO THINK 2

Would you describe bacteria as being helpful or harmful to people? Explain.

Figure 2: Some Common Types of Bacteria

Shape	Examples	Ecological Roles
sphere 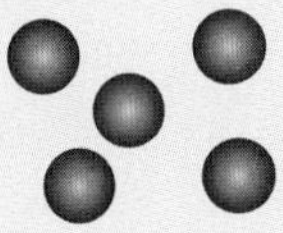	*Diplococci* (pairs of cocci) 	cause pneumonia
	Staphylococci (clusters of cocci) 	are normally present on human skin; some cause boils and infections
	Streptococci (chains of cocci) 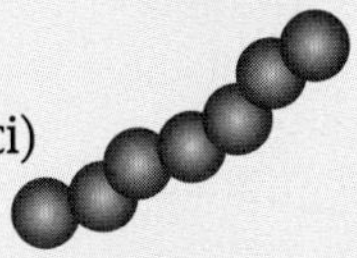	are used to make yogurt and cheese; cause strep throat
rod 	*Bacilli* (rods)	decompose hay; are used to make cheese, yogurt, pickles, and sauerkraut; are normally present in the human digestive tract; cause diarrhea; cause anthrax in cattle and sheep
	Mycobacteria (chains of bacilli) 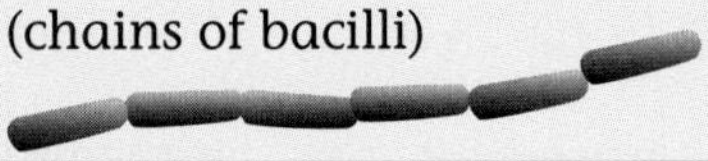	cause tuberculosis; are found normally in soil and water.
curved rod 	*Vibrio*	cause cholera; help break down sewage
short spirals 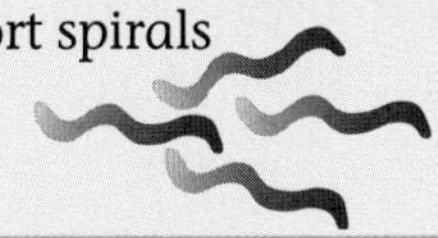	*Spirilla*	are decomposers in both fresh and salt water
long spirals 	*Spirochete*	cause syphilis; are decomposers
branched chain 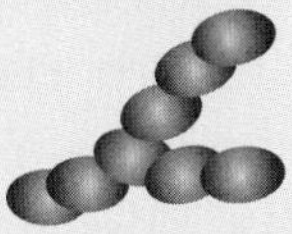	*Actinomyces*	produce several antibiotics; were once classified as fungi

Cocci are spherical bacteria; the singular of cocci is coccus.

Viruses: A Group Apart

Viruses are not living organisms. Unlike protists, bacteria, and all other living organisms, viruses are not made up of cells. They are unable to grow or reproduce independently or carry out the functions, such as respiration, that living organisms do. Instead, viruses rely on the cells of living organisms for their reproduction. It is for this reason that we say infectious diseases are caused by microbes, and not microorganisms.

Figure 3: Comparing Average Sizes of Microbes
These are relative, not actual, sizes of microbes. An average bacterium is actually much smaller than the virus shown here.

STOPPING TO THINK 3

a. Why are viruses not considered to be microorganisms?

b. Look at Figure 3, "Relative Sizes of Microbes." How do the sizes of protists, bacteria, and viruses compare?

c. Which do you think cannot be seen with a classroom microscope?

How do we know viruses exist? The existence of viruses was first suggested in 1898, nearly 45 years before they were first seen. In 1895, Dutch scientist Martinus Beijerinck (BY-er-ink) began experimenting with the tobacco plant. He was studying a plant disease that he believed to be infectious. By this time, scientists were familiar with protists and bacteria, so Beijerinck began searching for a bacterium that might be causing this disease. But he could not find one. Yet his experiments demonstrated that the disease could be passed from plant to plant, so he concluded that the disease was caused by a microbe. Since it wasn't a protist or a bacterium, he called it a virus, which means "poison" in Latin.

Viruses are so small that you need an electron microscope to see one. The electron microscope was not invented until the 1930s. As a result, viruses were first seen in 1939. Today, we know that viruses cause many diseases, including the flu, colds, chickenpox, and AIDS.

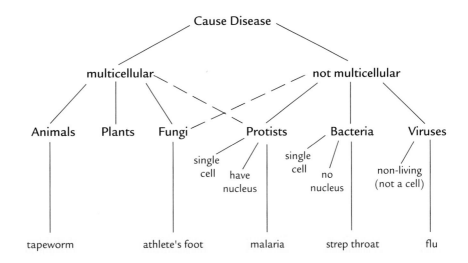

Figure 4: Classifying Disease-Causing Organisms and Viruses

Figure 4 shows the five-kingdom classification plus viruses. Note the examples of diseases caused by members of each group. What do you think the dotted lines mean?

For links to more information about microbes, go the SALI page of the SEPUP website.

ANALYSIS

1. You have read how microbes can be both helpful and harmful to humans. Do you think a microbe can be *neither* helpful nor harmful? Explain.

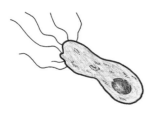

2. You decide to examine some pond water under a microscope. With a magnification of 40 (using the 4x objective), you observe a long, cylindrical organism moving across your field of view (see left). As you look more closely, you notice what appears to be a round structure inside of it. Is this organism most likely a protist, bacterium, or virus? Explain how you arrived at your conclusion.

3. Suppose your school's microscopes did not have 40x objectives, but only 10x objectives. Your friend, who is in high school, uses a 40x objective. Explain what group of microbes he or she can study that you cannot.

4. What are the advantages of using the highest power objective on a microscope? What are the advantages of using the lowest power objective on a microscope? Explain.

5. In your science notebook, draw a larger version of the Venn diagram shown below. Record unique features of each group of microbes in the appropriate space. Record common features among groups in the spaces that overlap. **Hint:** Think about what you have learned about cells in the last few activities.

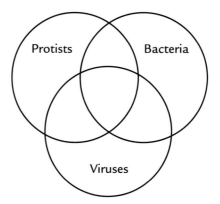

6. **Reflection:** On a field trip, you visit a laboratory that has an electron microscope. The microscopist (the person who runs the microscope) offers to set up a microbe for you to view. What microbe, or group of microbes, would you choose to view? Why?

46 Disease Fighters

INVESTIGATION

What does your body do to protect itself from invading microbes? Even before an organism can enter your body, your skin provides a protective barrier. But foreign substances can still enter through cuts or natural body openings, such as your mouth or your nose. Tears, saliva, and mucus help to remove some invaders at these sites. But when foreign substances cross these barriers, your **immune** (ih-MYOON) **system** comes to the rescue.

Your immune system has the amazing ability to distinguish between the substances of your own body and foreign substances, such as bacteria and viruses. A healthy immune system can then mount a defense against these invaders. Several kinds of cells, particularly white blood cells, are responsible for this immune response. The pictures here show normal human blood cells. Note that the red blood cells are the most common. Also note the detail of the white blood cells. They increase in number when the body is under attack from a foreign substance.

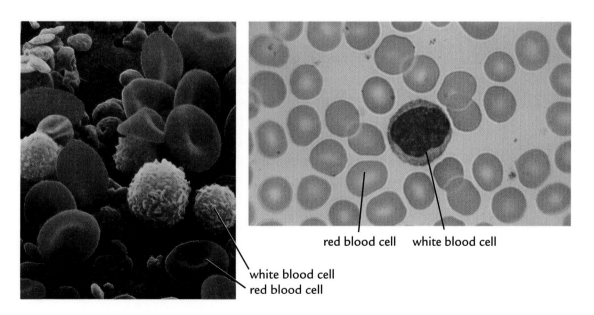

The photograph on the left was taken through a scanning electron microscope, while the photograph on the right was taken through a light microscope.

Activity 46 • Disease Fighters

Immune responses of the human body are not always helpful. Any new material in the body, including blood and organs, can trigger an immune response. It is this reaction of the immune system that makes organ transplants and blood transfusions difficult. If the blood type of the blood donor is not compatible with that of the person receiving the blood, the transfused blood cells are seen as foreign by the immune system and they clump together. These clumps can create blockages in blood vessels and cause death. That is why it's important to know which types of blood can be donated safely to people with each of the four human **blood types: A, B, AB, and O.** You will simulate what happens to a person's blood when blood from a donor is added.

For links to more information on the blood and diseases of the blood, go the SALI page of the SEPUP website.

How does your blood help fight infectious diseases?

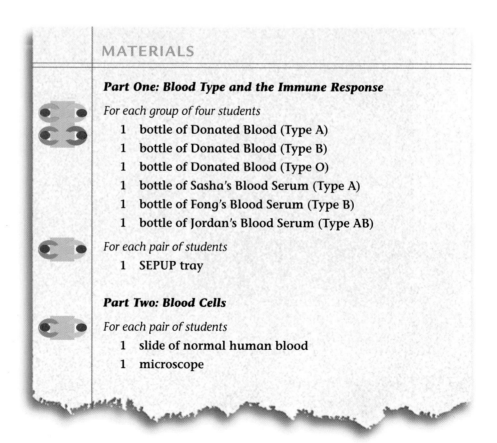

MATERIALS

Part One: Blood Type and the Immune Response

For each group of four students
- 1 bottle of Donated Blood (Type A)
- 1 bottle of Donated Blood (Type B)
- 1 bottle of Donated Blood (Type O)
- 1 bottle of Sasha's Blood Serum (Type A)
- 1 bottle of Fong's Blood Serum (Type B)
- 1 bottle of Jordan's Blood Serum (Type AB)

For each pair of students
- 1 SEPUP tray

Part Two: Blood Cells

For each pair of students
- 1 slide of normal human blood
- 1 microscope

PROCEDURE

Within each group of four students, one pair begins with Part One and the other pair begins with Part Two. When both pairs have completed their parts, they can switch roles.

Part One: Blood Type and the Immune Response

BLOOD EMERGENCY!

Three patients needing blood transfusions have arrived at the local hospital. This is the chart showing their blood types. In order to supply the blood, the hospital staff has asked the community to help. Several people respond by donating blood. The hospital receives blood donations of types A, B, and O, but these blood types might not be compatible with each patient.

Patient	Blood Type
Sasha	A
Fong	B
Jordan	AB

Does the hospital have enough of the right type of blood for each patient? Find out by testing samples of each blood type.

1. Collect the three blood samples and the three serum samples.

   Note: *Serum is blood that has had the red blood cells removed. In blood transfusions, the donor's blood must be compatible with the patient's serum.*

2. Design a data table to record your experimental results. You will test each of the three donated blood types with serum from each of the three patients.

3. Place two drops of Sasha's Blood Serum in Cups 1–3 of your SEPUP tray.

Activity 46 • Disease Fighters

4. Add two drops of Donated Blood Type A to Cup 1. Record the results in your data table.

5. Test Sasha's Blood Serum with the remaining donated blood samples. Record the results in your data table.

6. Use Cups 4–9 to test the samples from the other two patients, Fong and Jordan. Record the results in your data table.

Part Two: Blood Cells

7. You and your partner should receive a microscope slide of normal human blood.

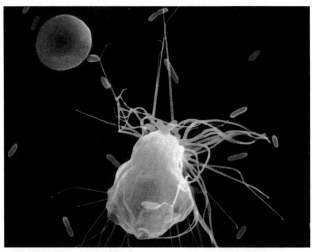

This high power scanning electron microscope photograph has been colorized. A red blood cell is near the top of the picture. A white blood cell (colored purple) is attacking bacteria (colored blue-green).

8. Be sure that your microscope is set on the lowest power (shortest objective) before placing your slide onto the microscope stage. Center the slide so that the specimen is directly over the light opening and adjust the microscope settings as necessary.

   Hint: To check that you are focused on the material that is on the slide, move the slide slightly while you look through the eyepiece—the material that you are focused on should move at the same time you move the slide.

9. Begin by observing the slide on low power (usually the 4x objective). Scan the slide and focus on a section that shows more than one kind of cell.

   Hint: Remember that stains are often used to make structures on a slide more visible. Look carefully for a light pink smear with a dark purple blob. If material on the slide is too light to see, reduce the amount of light on the slide: do this by slightly closing the diaphragm under the stage.

10. Without moving the slide (which can be secured with stage clips), switch to medium power (usually 10x). Adjust the microscope settings as necessary.

11. Without moving the slide, switch to high power (usually the 40x objective). *Be careful not to smash the objective against the slide!* Adjust the microscope settings as necessary.

Hint: If material on the slide is too dark to see, increase the amount of light on the slide: do this by slightly opening the diaphragm under the stage.

12. In your science notebook, describe the two different kinds of cells that you see. In your description, include which type of cell is more common, the shape of each cell, the relative size, and any cell structures you are able to identify in either cell.

ANALYSIS

Part One: Blood Type and the Immune Response

1. Each patient required one pint of blood. The hospital received one pint each of type A, B, and O blood. Explain whether the hospital had enough of the right type of blood for each patient.

2. What prevents your body from accepting transfusions of certain types of blood?

Part Two: Blood Cells

3. Think back to all the work that you have been doing on cells. Compare and contrast different types of cells by copying and completing the table below.

4. In what ways does your body prevent you from catching an infectious disease?

Cell Type	Cell Shape	Cell Membrane?	Cytoplasm?	Nucleus?
Bacteria				
Protist				
Plant (onion)				
Animal: cheek				
Animal: red blood cell				
Animal: white blood cell				

47 Reducing Risk

How do you prevent yourself from catching an infectious disease? You now know that your immune system provides you with natural defenses, but sometimes your immune system becomes overwhelmed by disease-causing microbes and you get sick. One way to reduce your risk of getting sick is by taking simple precautions, like washing your hands before you eat. You may even use antimicrobial solutions, such as an antibacterial soap or a disinfectant, when you clean up. How effective are these products at killing germs?

You can measure the effect of different solutions on the growth of microbes. You can culture, or grow, microbes in a special dish known as a **petri** (PEE-tree) **dish**. The petri dish contains food for the organisms you are trying to culture. In classrooms, the most commonly used food is agar, a gelatin-like material that was first invented by Robert Koch. If bacteria are present, many of them will grow on the agar. You can see this in the petri dish in the photograph below. It is also possible for molds and algae to grow on agar.

How effective are different solutions at preventing the growth of microbes?

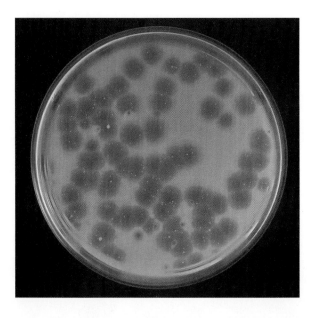

Reducing Risk • Activity 47

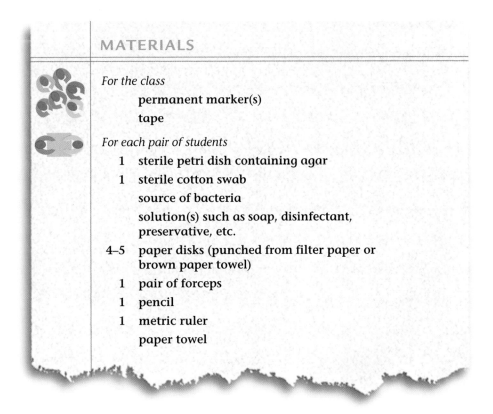

MATERIALS

For the class
- permanent marker(s)
- tape

For each pair of students
- 1 sterile petri dish containing agar
- 1 sterile cotton swab
- source of bacteria
- solution(s) such as soap, disinfectant, preservative, etc.
- 4–5 paper disks (punched from filter paper or brown paper towel)
- 1 pair of forceps
- 1 pencil
- 1 metric ruler
- paper towel

PROCEDURE

1. Use a permanent marker to put your name and class period on the bottom of your petri dish.

2. Dip your swab in the source of bacteria provided by your teacher.

3. Remove the lid of the petri dish. Use the swab to streak the bacteria onto agar as shown in Figure 1. Press firmly but not too hard. Your goal is to spread the bacteria but not to break up the agar layer. Turn the dish 90° and repeat as shown in Figure 2.

Figure 1: First Streak   **Figure 2: Second Streak**

C-83

4. Use a pencil to write the initial of the solution you are testing on a disk of filter paper (or brown paper towel).

5. Use your forceps to dip the disk in the solution. To remove excess solution, touch the edge of the disk to a clean paper towel.

6. Place the disk on the agar of your petri dish and re-cover the dish.

7. Repeat Steps 3–6 for each solution you are testing. Be sure to leave at least 1.5 cm between the paper disks. Tape around the dish to seal it.

8. Place your petri dish in a warm place for a few days.

9. Check the growth of microbes in the dish each day. Record if the growth of microbes has stopped in the area around the disk. If so, use a ruler to measure this space, known as the zone of inhibition.

10. Examine the control dishes set up by your teacher. Record your observations.

ANALYSIS

1. Did the solution(s) affect the growth of bacteria on your petri dish? Explain. Be sure to compare your results to the control and to describe your evidence.

2. Share your results with the class.

    a. Did everyone who tested the same solution get the same results? Explain.

    b. How effective were the different solutions in preventing microbial growth?

    c. How might you follow up on this investigation or improve the design of this investigation?

3. **Reflection:** Have the results of your experiment caused you to want to change any of your behavior (such as what solutions you use to wash your hands or household surfaces)? Why or why not?

48 Wash Your Hands, Please!

INVESTIGATION

In the 1840s, Dr. Semmelweiss found that hand washing could significantly reduce the rate of infection in hospitals. One common type of illness that can be reduced by hand washing is food poisoning. Millions of people suffer from some form of food poisoning each year. Most people who get these infections don't die, but they feel terrible and miss work or school days.

Could hand washing reduce the number of times you get sick? Hands have about 200 million microbes on them. Most are harmless, but some of these microbes can cause food poisoning, colds, flu, and other infections. In fact, public health researchers estimate that 80% of common infections in the U.S. are caught by touching surfaces that are contaminated with infectious microbes. Contaminated surfaces might include sinks, countertops, doorknobs, or your own hands.

CHALLENGE

How effectively does hand washing reduce the spread of microbes? How can you improve the effectiveness of hand washing?

Activity 48 • Wash Your Hands, Please!

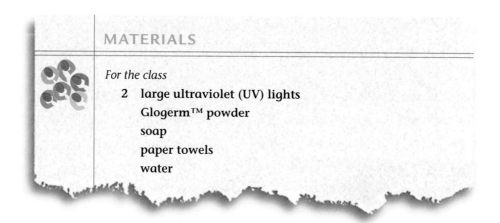

MATERIALS

For the class
2 large ultraviolet (UV) lights
Glogerm™ powder
soap
paper towels
water

PROCEDURE

Part One: Washing Your Hands

1. In your science notebook, make a table like the one shown below.

Table 1: Observations of Hands

	Hands Sprinkled With Powder	Hands That Were Shaken
Before		
After		

2. Have one person on your team of two students sprinkle a small amount of white powder on the palm of one hand. This person should spread the powder all over his or her hands by rubbing the hands together, covering the palms, backs of hands, fingers, and nails.

3. Firmly shake both hands with your partner. Do this by shaking right hand with right hand and left hand with left hand.

4. Look carefully at your hands and your partner's hands under the ultraviolet (UV) light. Record your observations in the first row of your data table.

5. Both you and your partner should wash your hands as you would *normally*.

6. Look carefully at your hands and your partner's hands under the UV light. Record your observations in Table 1.

Part Two: Improving Hand Washing

7. Design an experiment to improve the effectiveness of hand washing in removing microbes from the surface of your hands. For example, does the length of time you rub your hands make a difference? Is there a specific technique that is better for hand washing?

When designing your experiment, think about the following questions:

- What is the purpose of your experiment?
- What variable are you testing?
- What variables will you keep the same?
- What is your hypothesis?
- How many trials will you conduct?
- Will you collect qualitative and/or quantitative data? How will these data help you make a conclusion?
- How will you record these data?

8. Record your hypothesis and your planned experimental procedure in your science notebook.

9. Make a data table that has space for all the data you need to record. You will fill it in during your experiment.

10. Obtain your teacher's approval of your experiment.

11. Conduct your experiment and record your results.

ANALYSIS

Part One: Washing Your Hands

1. Where on your hands did you find the most "microbes" (white powder)?

2. Based on your results in Part One, how well did washing your hands remove "microbes"?

Part Two: Improving Hand Washing

3. Why is hand washing important? Use your knowledge of microbes and the results of this activity to explain your answer.

4. How well do powdered "microbes" model real microbes? Explain.

5. Imagine that your school has decided to launch a hand-washing campaign. You are in charge of designing the campaign and evaluating its effectiveness.

    a. Why might people resist changing the frequency and the way in which they wash their hands?

    b. Explain how you could persuade people to change their hand-washing behavior.

    c. What type of data could you collect (both before and after the campaign) to determine if the hand-washing campaign was effective?

6. Read the recommendations for hand washing for surgeons and food handlers on the next page. Why do both sets of guidelines stress rubbing or scrubbing the hands?

EXTENSION

Make a list of recommendations for a school hand-washing campaign. Explain how each recommendation would help reduce the spread of microbes.

Guidelines for Doctors Prior to Surgery

- Wet hands.

- Clean nails.

- Scrub hands (fronts and backs, each finger and between fingers) and forearms for 5 minutes, using antibacterial soap and a hand brush.

- Hold hands above elbow and allow excess water to drip off.

- Dry hands and forearms with a sterile towel.

- Put on surgical gown.

- Put on sterile gloves. (Many surgeons use double gloves.)

Guidelines for Food Industry Workers
(restaurant staff, supermarket workers, food packers, etc.)

- All personnel must wash their hands before returning to work.

- Wet hands with warm running water.

- Add soap, then rub hands together, making a soapy lather. Do this away from the running water for at least 15 seconds, being careful not to wash the lather away.

- Wash the front and back of hands, as well as between fingers and under nails.

- Rinse hands well under warm running water. Let the water run back into the sink, not down to your elbows.

- Turn off the water with a paper towel and dispose in a proper receptacle.

- Dry hands thoroughly with a clean towel.

49 An Ounce of Prevention

ROLE PLAY

As you first learned in Activity 46, "Disease Fighters," your immune system recognizes and fights disease-causing microbes. Most people are able to fight off diseases like colds or the flu quickly and return to full health within a week or so. Other diseases, like diphtheria, are more severe. Such diseases are much more likely to have serious effects, or even lead to death, in a larger portion of the population.

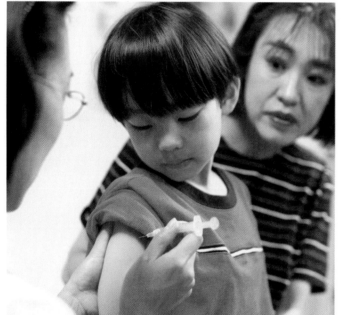

How can you fight serious diseases that often overwhelm human immune systems? One approach, **vaccination** (vak-suh-NAY-shun), is very effective in preventing some diseases. Each **vaccine** works against a specific disease. Vaccines are available against diseases caused by both viruses and bacteria. There are also vaccines being developed to work against other microbes as well.

CHALLENGE

How do vaccines prevent disease?

PROCEDURE

1. Assign a role for each person in your group. Assuming that there are four people in your group, each of you will read one role in Scene 1 and another role in Scene 2.

Roles in Scene 1	Roles in Scene 2
Student	Student
Parent	Parent
Older sibling (sister or brother)	Doctor
Grandparent	Nurse

2. Read the role play on the next pages aloud. Insert the names of your group members as directed.

ACT OR REACT!

SCENE 1: At the dinner table

Parent: It's getting to be flu season. My boss wants all employees to get flu shots. She doesn't want us to miss work and get behind on our deadlines.

Older Sibling: Funny you should mention that. My friend at school was just telling me that he has been feeling sick and thinks he has the flu. How does a flu shot stop you from getting sick?

Grandparent: A flu shot is a vaccine that helps your body resist the flu.

Student: Oh, I know about vaccines. I had a measles vaccine when I was a little kid and then you took me to get a booster shot just last year.

Parent: That reminds me—it's time for your tetanus booster.

Student: Another shot? I hate shots.

Parent: Yes, the immunity from some vaccines begins to wear off after a while, so you need a booster—it gives your immune system a boost.

Student: Why do I need to get all of these shots? And how does a vaccine work anyway?

Older Sibling: A vaccine is a dead or weakened form of a microbe or a part of the microbe. It helps your immune system prepare in advance to fight off the disease-causing germs.

Student: You mean they actually inject you with the disease microbe?

Older Sibling: Yup! The microbe is first inactivated, so it doesn't make you sick.

Student: Do vaccines work only against diseases caused by viruses, like the flu?

Grandparent: I don't think so. There are vaccines against tetanus and diphtheria, and they are caused by bacteria, not viruses. When my mother and father were very young, their parents worried that they would get tetanus, or "lockjaw," as they called it, every time they got a deep cut. They didn't have tetanus shots then.

Student: So you're saying that all vaccines work to keep you from getting a disease, but not all diseases are caused by the same thing.

Parent: Exactly. The way in which vaccines work is the same for different diseases, but each disease is caused by a different microbe, or germ. A microbe can be a bacterium or a virus.

Student: Then what's the difference between bacteria and viruses?

Older Sibling: This is exactly what we're studying in science class! Bacteria are living cells that grow and divide. Viruses can't grow or divide unless they inject their genetic material into another cell.

Grandparent: (Name of Older Sibling), what do you mean by "genetic material"?

Older Sibling: Oh, that just means DNA. You've probably heard of DNA.

Student: I've heard of DNA. Viruses have DNA?

Parent: Yes, I read about this. Viruses have a small amount of DNA, or sometimes a similar substance called RNA, as their genetic material. They also have an outer coat that protects the material inside, but that's about it. They infect cells and cause the infected cells to make new copies of the virus. But if a virus can't get into a living cell, no copies are made and the virus can't reproduce. That's why viruses aren't considered to be living organisms.

Student: How do they make a vaccine? Why doesn't it make you sick?

Grandparent: I don't know about all vaccines, but I do remember polio vaccines. Polio was a common disease when I was young. In fact, President Franklin D. Roosevelt became paralyzed from the waist down when he caught polio as an adult. He hid from the public how serious it was, perhaps because of people's attitudes toward disabilities at the time.

Chemicals or heat were used to inactivate the polio virus and it was then injected into a healthy person. As (*Name of Older Sibling*) said, the inactivated virus didn't make you sick. But your immune system was tricked into getting ready to fight off an infectious polio virus.

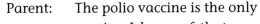
President Franklin Delano Roosevelt

Parent: The polio vaccine is the only vaccine I know of that you can take orally. They used to put it on a sugar cube. The weakened virus wouldn't make you sick, but would still cause you to become immune.

Grandparent: I remember the first polio vaccine. It was a shot. My little brother hated it, but my mother was so relieved not to have to worry that we might get polio. It was a very serious threat back then. Vaccines have almost wiped out some diseases like polio and smallpox.

Student: Wow, until today, I never thought about life before vaccines.

SCENE 2: In the doctor's office

Nurse: Dr. (*Last name of Doctor*), here's your next patient.

Doctor: Hello (*Name of Student*), what seems to be the problem? Not feeling well?

Student: I think I have the flu.

Doctor: (checking pulse) Hmmm…let's see. What are your symptoms?

Student: Well, I've been coughing a lot and my throat's sore.…I've been really tired. And I'm also starting to have a hard time breathing.

Parent: I think there's a fever, too.

Doctor: (to Student) Open your mouth and let's take a look.

(Doctor examines Student's throat, listens to chest cough, and takes temperature.)

Well, it could be the flu. But I'm not sure. You might have strep throat or pneumonia (new-MOW-nyah) or just a really bad cold. We'll have to take a chest x-ray and a throat culture.

Parent: Are you going to prescribe an **antibiotic** (an-tih-by-AH-tik)?

Doctor: Not yet.

Parent: Why not?

Nurse: Because if you have the flu, an antibiotic won't work. The flu is a viral disease, as are some types of pneumonia.

Parent: I'm sorry, but I don't understand. I thought it was standard practice to prescribe an antibiotic. I always get one when I have the flu or a sore throat.

Doctor: Nurse (*Name of Nurse*), why don't you explain while I give (*Name of Student*) a chest x-ray and throat culture?

(Student leaves with doctor.)

Nurse: Good idea. Antibiotics are medications that are used to fight bacterial diseases. This is because bacteria are living organisms that are killed by the action of an antibiotic.

Parent: They always tell you to take the entire prescription.

Nurse: Right. It takes time to kill the entire population of bacteria in your body. If you stop taking the antibiotic before all the disease-causing bacteria are gone, you run the risk of having them increase again.

Parent: So why don't antibiotics work on diseases caused by viruses?

Nurse: Because of the fact that viruses are not cells. They can reproduce only by entering your cells and using the cells to reproduce. That makes them harder to attack....Oh, here's (*Name of Student*), back from the x-ray room.

(Doctor and Student return.)

Student: Hey (*Mom/Dad*), my X-ray was negative, so I don't have pneumonia. I hope I don't have to take antibiotics. I hate swallowing pills!

Parent: Doctor, why don't you just go ahead and give us an antibiotic? The nurse just said that antibiotics are the way to treat bacterial diseases.

Doctor: That's right, but unless a throat culture or X-ray is positive, (*Name of Student*) probably doesn't have a bacterial infection. And it isn't helpful to take antibiotics that are not needed.

Student: But the last time I had the flu, you prescribed an antibiotic.

Doctor: That's right. But if I recall correctly, the last time you were here, your little sister was also sick—and she had a bacterial infection. You seemed to have the early symptoms of her infection. So the antibiotics were intended to treat the bacterial infection.

Student: (interrupting)...and not to cure the flu! So how do you cure the flu?

Nurse: There is no cure for the flu. Medicines you take when you have the flu or a cold only relieve the symptoms, like fever or headache. But they can't make you well; they can just make you feel better. In fact, there is no cure for most viral diseases.

Parent: What about flu shots?

Student: (*Mom/Dad*), you told me that flu shots prevent you from getting the flu. They don't make you better.

Doctor: That's right. Unfortunately, although flu shots prevent the flu in most cases, they don't work in every case. Why, just this morning we had a lady in the office who had gotten a flu shot, but who still caught the flu. It seems that she was infected with a different type of the flu than the one she had received a vaccine for.

Nurse: May I suggest that you take (*Name of Student*) home? We'll call you in the morning; by then we'll know if (*Name of Student*) needs an antibiotic.

Student: Yeah, I need to lie down. My head hurts from all this stuff. Maybe I'll ask my science teacher more when I feel better.

ANALYSIS

1. A vaccine prevents a person from catching an infectious disease; it does not treat the disease after the person has caught it. What are some advantages of preventing, rather than treating, infectious diseases?

2. Why are serious side effects from vaccines very rare?

3. You go to the doctor and find out that you may have the flu. Would you expect to be prescribed an antibiotic? Explain your answer.

4. Do you think that vaccinations against the flu should be required? Explain. Support your answer with evidence and identify the trade-offs of your decision.

    Hint: To write a complete answer, first state your opinion. Provide two or more pieces of evidence that support your opinion. Then consider all sides of the issue and identify the trade-offs of your decision.

5. **Reflection:** Explain whether you would change your answer to Question 4 if the disease had more severe symptoms and a greater chance of causing death.

EXTENSION

The vaccines for polio were developed in the 1950s. Find out more about how this disease affected society by asking different generations of your family, such as your parents and grandparents, if they can recall knowing anyone who had polio.

50 Fighting Back

VIEW AND REFLECT

Despite many prevention methods, infectious diseases, from Hansen's disease to tuberculosis, continue to affect people in the U.S. and around the world. What can be done for a person after he or she has caught an infectious disease?

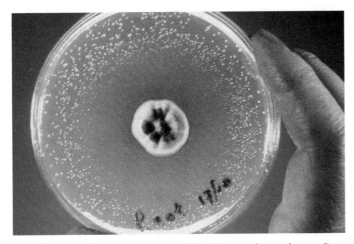

Penicillium *mold produces a chemical that has saved many human lives.*

Diseases caused by microorganisms, such as bacteria and protists, can usually be treated with antibiotics. This is because antibiotics are chemicals that kill living microbes such as bacteria. You may have heard of antibiotics such as penicillin (peh-nuh-SIH-lun) and streptomycin (strep-tuh-MY-sun). You may have even used them yourself. How were antibiotics first discovered? Did scientists design experiments and control variables? What problems did scientists face?

CHALLENGE

How was the first antibiotic discovered?

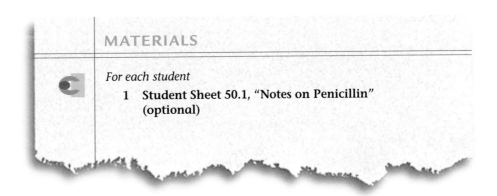

MATERIALS

For each student

1 Student Sheet 50.1, "Notes on Penicillin" (optional)

Activity 50 • Fighting Back

PROCEDURE

1. In order to prepare to watch the story on the video, first read Analysis Questions 1–3.

2. Find out more about antibiotics by watching a segment on the discovery of penicillin from the video, *A Science Odyssey:* "Matters of Life and Death."

3. Watch the video a second time and take notes on the following questions. Or use Student Sheet 50.1, "Notes on Penicillin," as a guide.

   • What was the scientific discovery? How was it made?

   • Who made it?

   • What was done as a result of this discovery?

4. Answer the Analysis Questions.

ANALYSIS

1. Describe the impact of penicillin on society.

2. Think back to the traditional scientific method, first discussed in Activity 1, "Solving Problems: Save Fred!" in Unit A, "Studying People Scientifically," of *Science and Life Issues.* Explain how the work of each of the following scientists did or did not resemble the traditional scientific method.

   a. Alexander Fleming

   b. Oxford University team (made up of 19 researchers, including Howard Florey and Ernst Chain)

3. What types of infectious diseases do antibiotics work against? Are there any types of infectious diseases that antibiotics do not work against? Explain.

51 The Full Course

MODELING

Have you ever taken antibiotics? Did you follow the directions completely? All antibiotics need to be taken as directed, which usually means taking all the pills and not stopping even if you begin feeling better. Why?

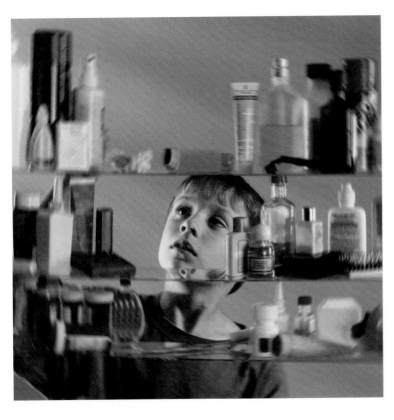

Millions of harmless bacteria naturally live on and inside of your body. When harmful bacteria appear on the scene, your body's immune system can usually keep a small population of them under control. If, however, these bacteria reproduce too quickly, you suffer consequences—and this is called an infection. Antibiotics help your body fight off an infection by killing these harmful bacteria. Unfortunately, a small number of bacteria in any population may not be affected by the antibiotic as quickly. These bacteria, which are considered more **resistant** to the treatment, continue to reproduce and grow. Completing the **full course** of the antibiotic as prescribed helps make sure that these bacteria do not survive and therefore won't make you ill or infect anyone else.

Why is it important to take an antibiotic as prescribed?

Activity 51 • The Full Course

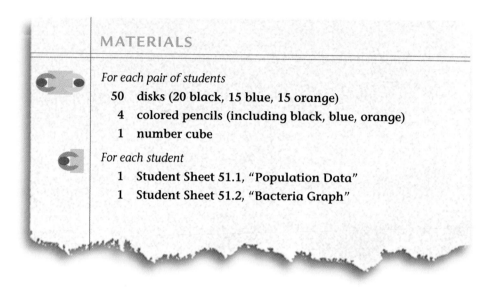

MATERIALS

For each pair of students
- 50 disks (20 black, 15 blue, 15 orange)
- 4 colored pencils (including black, blue, orange)
- 1 number cube

For each student
- 1 Student Sheet 51.1, "Population Data"
- 1 Student Sheet 51.2, "Bacteria Graph"

A BACTERIAL INFECTION

Imagine that you are sick with a bacterial infection. Your doctor prescribes an antibiotic to be taken every day for eight days.

Colored disks represent the harmful bacteria that are in your body:

Disease-Causing Bacteria	Represented By
Least resistant bacteria	black disks
Resistant bacteria	blue disks
Extremely resistant bacteria	orange disks

Each time you toss a number cube, it is time to take the antibiotic. The number on the number cube tells you what to do.

PROCEDURE

1. In this activity, you will work with your partner to collect data. Begin with 20 disks: 13 black, 6 blue, and 1 orange. These disks represent the harmful bacteria living in your body before you begin to take the antibiotic. Set the extra disks aside for now.

2. It is time to take your antibiotic. Toss a number cube and follow the directions in the Number Cube Key (on the next page).

Number Cube Key		
You Toss	**What Happened**	**What To Do**
1, 3, 5, 6	You took the antibiotic on time, so bacteria are being killed!	Remove 5 disks: remove all of the black disks first, then the blue, then the orange.
2, 4	You forgot to take the antibiotic.	Do nothing.

3. Record the number of each type of bacteria in your body in Table 1, "Number of Harmful Bacteria in Your Body," on Student Sheet 51.1, "Population Data."

4. *The bacteria are reproducing all of the time!* If one or more bacteria of a particular type are still alive in your body, add 1 disk of that color to your population.

   For example, if you have resistant (blue) and extremely resistant (orange) bacteria in your body, add 1 blue disk and 1 orange disk to your population.

5. Repeat Steps 2–4 until you have completed Table 1.

6. Use your data in Table 1 to graph the population for each type of bacteria and for the total number of bacteria on Student Sheet 51.2, "Bacteria Graph." Use different colored lines, or lines with different patterns, to represent each type of bacteria, and fill in the key accordingly.

Activity 51 • The Full Course

ANALYSIS

1. Did the antibiotic help you to completely kill all of the harmful bacteria living in your body? Explain.

2. **a.** Imagine infecting someone else immediately after catching the infection (before you started taking the antibiotic). With what type of bacteria would you be most likely to infect them?

   b. Imagine infecting someone else near the end of your antibiotic course. With what type of bacteria would you be most likely to infect them?

   c. Suppose most infected people stopped taking the antibiotic when they began to feel better. (For example, consider the point in the simulation when there were only three harmful bacteria left.) What do you predict might happen to an antibiotic's ability to kill the harmful bacteria if the infection returns? Explain your reasoning.

3. Use your graph to describe how the population of each type of bacteria changed over the course of the antibiotic treatment.

4. Why is it important to complete the full course of an antibiotic as prescribed?

5. Was this activity a good model of an antibiotic treatment? Explain.

6. You find out that you have a viral infection and not a bacterial infection. What would happen to the amount of virus in your body each time you took the antibiotic? Explain.

52 Miracle Drugs—Or Not?

TALKING IT OVER

In the last activity, you saw how important it is to follow directions and complete the full course of antibiotics as prescribed. Are antibiotics truly miracle drugs? Will they cure every infection? What can people do to maintain the effectiveness of antibiotics?

CHALLENGE

What happens when antibiotics are overprescribed or used improperly?

PROCEDURE

1. Read about the miracle drugs known as antibiotics. As you read, think about what you might do if you were prescribed an antibiotic.

2. Discuss Analysis Question 1 with your group.

MIRACLE DRUGS—OR NOT?

What if someone told you that the pill you took to get better today might not work for you if you fall sick tomorrow? That's what health experts are saying about the miracle drugs known as antibiotics.

Common Antibiotics

Antibiotic	Brand Name	Used Against
Amoxicillin	Amoxil®, Polymox®, Wymox®, Trimox®	bronchitis, ear infections, sinus infections
Ampicillin	Unasyn®	urinary tract infections, meningitis
Cefaclor	Ceclor®	infections of the ear, nose, throat, respiratory tract, and urinary tract; strep throat; pneumonia; tonsillitis
Ceftriaxone	Rocephin®	Lyme disease, gonorrhea
Cephalexin	Keflex®, Keftab®	infections of the skin and urinary tract
Chloramphenicol	Chloromycetin®	typhiod, Rocky Mountain Spotted fever, meningitis
Clotrimazole	Lotrimin®, Mycelex®	yeast infections
Clindamycin	Cleocin®	pneumonia, strep throat, acne
Doxycycline	Atridox®, Doryx®, Doxy®, Periostat®, Vibramycin®	urinary tract infections, chlamydia, trichomonas
Erythromycin	Akne-Mycin®, EryDerm®, Erygel®, Ery-Tab®, Erythrocin®, Ilotycin®, Staticin®	Legionnaire's disease, pneumonia, strep throat, mild skin infections
Isoniazid	Nydrazid®	tuberculosis
Metronidazole	Flagyl®	amoebic dysentery, giardiasis
Monocycline	Minocin®	acne, amoebic dysentery, anthrax, cholera, plague, respiratory infections
Mupirocin	Bactroban®	skin infections, impetigo
Penicillin	various	strep throat, pneumonia, syphilis, dental and heart infections
Tetracycline	Achromycin®	respiratory infections, pink eye, pneumonia, severe acne, typhoid, Rocky Mountain Spotted fever

Antibiotics have been used to fight diseases for over 50 years. Today, they are losing their effectiveness. This is the result of more antibiotic-resistant bacteria. In the last 10–15 years, antibiotic-resistant bacteria have included strains of *Mycobacterium tuberculosis*, which causes tuberculosis (TB), and *Streptococcus pneumoniae*, the most common cause of human ear and sinus infections.

Shown on page C-104 are some common antibiotics. Do you recognize any antibiotics that you have taken?

Reasons for the development of antibiotic-resistant bacteria include overprescription and incorrect use of antibiotics. "Using antibiotics incorrectly has led to the development of bacteria that can resist them," says Dr. Richard Dietrich of Kaiser Permanente in Baltimore, Maryland. This means that the drugs people rely on to cure everything from strep throat to bacterial pneumonia may not work when they are taken.

Colony of bacteria. Some microbes are antibiotic-resistant (orange).

Use of antibiotics kills majority of bacteria, except those that are antibiotic-resistant.

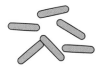

Without competition, antibiotic-resistant bacteria increase.

Figure 1: The Rise of Antibiotic-Resistant Bacteria

Most antibiotics must be taken over a period of time. When patients feel better, they sometimes stop taking the medication and don't complete the full course of treatment. In such cases, antibiotic-resistant bacteria may not be killed by the medication. They are more likely to reproduce and grow without competition from other microbes that have been killed by the drug. If the antibiotic-resistant bacteria cause disease, it becomes difficult to treat the patient with antibiotics (see Figure 1).

"The bacteria that cause pneumonia and ear and sinus infections commonly live in our throats and noses," Dr. Dietrich says. "If you take an antibiotic for no good reason, it kills only the germs that are not resistant to the antibiotic. An infection caused by the remaining resistant bacteria can be very hard to treat." If you take antibiotics when you don't need them, the drugs may lose their ability to help you get better when you really do need them.

One reason antibiotics are overused is that so many patients ask for them. Dr. Dietrich adds that it's common for patients to believe that antibiotics will cure whatever illness they have. But antibiotics work against only certain microbes, such as bacteria. They do not work against viruses. Doctors also used to prescribe antibiotics more often, partly as a precaution against disease. Now, this is less common and antibiotics are prescribed only for specific diseases.

Adapted with permission from Kaiser Permanente, 2001

Activity 52 • Miracle Drugs—Or Not?

 For links to more information on antibiotic resistance, go to the SALI page of the SEPUP website.

ANALYSIS

 1. Describe what can happen if people take antibiotics when they don't need them.

2. What is one reason antibiotics are overused?

 3. You have a sore throat and there are some antibiotics left over from your brother's strep infection last month. Should you take them for your sore throat? Why or why not?

 4. Your friend is prescribed an antibiotic on Monday. Suppose she feels better two days later. Should she stop taking the medicine? Explain.

5. **Reflection:** Think about what you have learned in the last few activities. Imagine you don't feel well and the doctor tells you that you have the flu. The doctor suggests taking an antibiotic. What would you do?

EXTENSION

Design a survey to find out what people know about the correct use of antibiotics. Good survey questions should be clear (the person should know exactly what you are asking) and concise (ask for only one piece of information per question). To make it easy to quantitatively analyze the survey data, develop questions that can be answered either yes or no. Examples of good questions include: When you are prescribed an antibiotic, do you always take all of it? Do you always expect a doctor to prescribe an antibiotic when you are sick?

53 Modern Outbreaks

INVESTIGATION

Ebola fever, Lassa fever, Hanta virus, Bolivian hemorrhagic fever, and AIDS are all examples of new infectious diseases. Some new infectious diseases are the result of new interactions between people and the environment. Many of these diseases can be traced to animal species. For example, when people go deeper into unexplored jungles, they are more likely to come into close contact with wild animals and their diseases, perhaps for the first time. In some cases, the disease passes from animal to human. Ebola is probably one such disease. Epidemiologists believe that Ebola originally may have been an infection of green monkeys in Uganda.

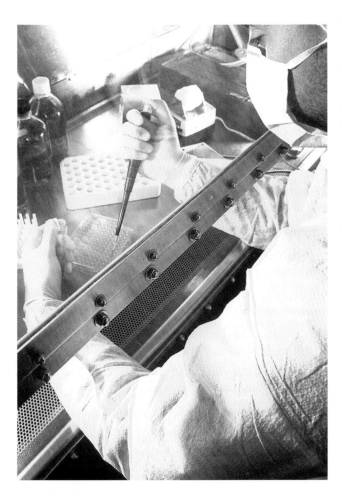

There is a real risk that such new diseases could quickly spread among different populations anywhere in the world. That is why the United States is prepared to send scientific teams to respond immediately to possible outbreaks of these new diseases, which are also known as **emerging diseases**.

In this activity, you will simulate the experience of a team of epidemiologists trying to trace the cause of a new disease.

Activity 53 • Modern Outbreaks

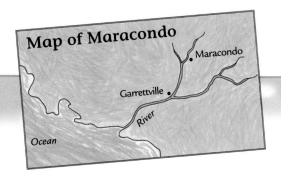

MARACONDO FEVER

An old man struggled out of a canoe in the tropical heat and stumbled into the town of Garrettville, collapsing in the main street. The children who found him were shocked to see that his eyes were bloodshot, his nose was bleeding, and his skin was bruised. He was immediately rushed to the hospital where he began coughing up blood. Before becoming unconscious, he told the doctors of a frightening disease affecting the people of his village, Maracondo. The old man died later that day of the same disease he had been warning the townspeople about.

None of the medical staff knew much about Maracondo (population 85), which is the last village that boats can reach as they head up the river. The medical staff of the Garrettville Hospital was worried about the spread of this mysterious disease. No one in the town became ill, but the townspeople were very frightened. They were also concerned about the people in Maracondo. The hospital doctors collected samples from the dead man, including urine, blood, and feces. They sent these samples, packed in dry ice, to the United States—to the Centers for Disease Control and Prevention (CDC) in Atlanta, Georgia. When the samples were analyzed, it appeared that the blood of the man contained an unknown virus.

The CDC worked quickly to put together an expert team to help the people of Maracondo. A doctor, an epidemiologist, a veterinarian, and an ecologist make up the team. The team's mission is to go to Maracondo to find out how this new disease is spread in order to stop more people from getting sick. If the illness is spread from person to person, it might take only one infected person to get on a plane to accidentally start an epidemic more horrifying and deadly than the bubonic plague of the 1300s.

• • •

You and your partners make up the expert team from the CDC. You must gather evidence to determine how this disease, now called Maracondo Fever, is transmitted. You need to do this in time to save the people of Maracondo.

As you begin your investigation, imagine you are heading into Maracondo. You are very hot, thirsty, and irritable. People in the village are terrified and tension is high. You have no idea if the village leader who meets you has the virus in his breath, or on his hands, or if the virus is being carried by mosquitoes that are, at this very moment, buzzing around your head. In addition, you will have to live in a straw hut, sleep in a hammock, and boil all of your water.

But as a "can do" person, you get local villagers to help you set up a lab in half the normal time. You also build an animal collection center to check if any of the local animals are carrying the disease. All the people on your team are experienced microbe hunters and know that hunches are not good enough. You owe it to the villagers and the rest of the world to base your conclusions and recommendations on strong evidence.

Modern Outbreaks • Activity 53

How is Maracondo Fever spread? What can you do to stop it?

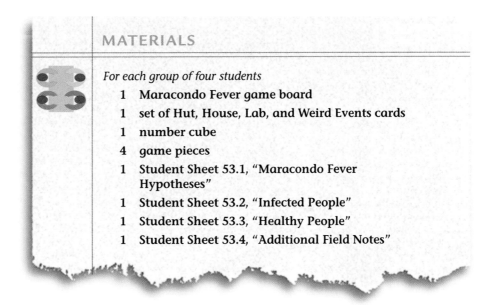

MATERIALS

For each group of four students
- 1 Maracondo Fever game board
- 1 set of Hut, House, Lab, and Weird Events cards
- 1 number cube
- 4 game pieces
- 1 Student Sheet 53.1, "Maracondo Fever Hypotheses"
- 1 Student Sheet 53.2, "Infected People"
- 1 Student Sheet 53.3, "Healthy People"
- 1 Student Sheet 53.4, "Additional Field Notes"

PROCEDURE

Part One: Field Notes

1. Before arriving in Maracondo, discuss with your team the symptoms and possible causes of the disease. Brainstorm ways in which the disease may be passed around the community.

2. After arriving at Maracondo, you gather additional information. Review your field notes (on the next page) before going on to Step 3.

3. How do you think Maracondo Fever is spread? Discuss your ideas in your group and write out your initial hypothesis on Student Sheet 53.1, "Maracondo Fever Hypotheses."

Field Notes: Maracondo May 25, 1963

After talking with local villagers, we have found that they have been growing more of their own food in the last two years. Areas of wild grassland and trees near the village have been cleared to grow corn, tomatoes, peppers, and other vegetables. Some villagers buy their food from river traders. The river traders do not eat local food.

The people of Maracondo live in two different types of homes: huts and small houses. Huts have dirt floors, while the houses have concrete floors. Depending on the family, people sleep on mats (on the floor) or on beds (off the floor). Mats do not usually have insect netting; beds may or may not have insect netting.

Animals that can be found in and around some of the homes include rats, mice, cats, dogs, bats, cows, and chickens. Insects, such as fleas and mosquitoes, are also present. Wild animals that live in the jungle nearby include monkeys and parrots.

A fiesta was held on March 28, 1963. By that time, some of our team had already arrived in Maracondo. Everyone helped prepare the food, which included corn meal (to make tortillas), corn on the cob, tomato and pepper salsa, beef, lemonade, and beer. Everyone enjoyed the fiesta, during which people shared food and drink.

Many of the town's 85 inhabitants have had Maracondo Fever. Some people have died, while others have been sick, but have since recovered. All recorded cases of sickness and death have been determined to be a result of the Maracondo Fever virus.

Part Two: Maracondo Fever Game

4. Begin the game by placing the cards face down in four stacks: Hut, House, Lab, and Weird Events. You will pick up a card every time you land on a space. For example, if you land on a Hut, read a Hut Card. Do the same for House, Lab, and Weird Events spaces.

5. Each person begins on the Start space. Have one person from the team toss the number cube and move that number of spaces on the game board.

6. Pick up and read the card. As a team, record the information you learn on

    - Student Sheet 53.2, "Infected People"
    - Student Sheet 53.3, "Healthy People"
    - Student Sheet 53.4, "Additional Field Notes"

7. As you gather more evidence, revise your hypothesis on Student Sheet 53.1, "Maracondo Fever Hypotheses." When you have too much evidence against a hypothesis, develop another hypothesis that fits the evidence.

8. Have the next player toss the number cube and move his or her game piece. As a team, repeat Steps 6 and 7.

9. Continue playing and collecting evidence. When the first person passes the Start space, pause to have a team meeting. Discuss how you think the disease was spread in light of the new evidence you have collected. Be sure to record your revised hypothesis on Student Sheet 53.1.

10. Continue playing and collecting evidence. When the next person passes the Start space, turn over all of the remaining cards and record all of the evidence.

11. As a group, complete Analysis Questions 1 and 2.

Activity 53 • Modern Outbreaks

ANALYSIS

1. **a.** Review your data on Student Sheet 53.2. What did the people who were infected have in common?

   b. Review your data on Student Sheet 53.3. What did the people who remained healthy have in common?

   c. Compare the data from Student Sheet 53.2 with the data on Student Sheet 53.3. What are some of the differences between those people who became infected compared with those who stayed healthy?

2. **a.** How do you think people are infected with Maracondo Fever? Explain how your evidence supports your final hypothesis.

   b. *People's lives are at stake!* Identify any evidence that seems to conflict with your final hypothesis and explain how your hypothesis addresses it.

3. Now that your CDC team has discovered how this disease spreads, you must recommend ways to reduce the spread of the disease, both within and outside of Maracondo. Recall what you know about viruses, as well as the information provided in this activity. Provide at least two recommendations to stop the spread of Maracondo Fever. Support them with evidence and identify the trade-offs involved in your recommendations.

   Hint: To write a complete answer, first state a recommendation. Provide two or more pieces of evidence that support your recommendation. Then consider the possible consequences of your recommendation and identify the trade-offs of your recommendation.

4. **Reflection:** What character traits and habits of mind would make a great epidemiologist?

Index

A **bold** page number identifies the page on which the term is defined.

A

actinomyces, C73
agar, C37, C82
AIDS, C60, C107
algae, C27, C28, C33
Amici, Giovanni, C22
Amoeba, C27, C46, C59, C72
animal cells, C34, C62–63
anthrax, C37, C73, C75
antibiotics, C20, C73, **C94**–95
    bacterial strains resistant to, C105
    brand names of, C104
    discovery of, C97–98
    full course of, **C99**–102
    improper use of/overprescribing, C103–106
    resistance to, C99
antimicrobial solutions, C82
antiseptics, C38
Archaea, C71
autopsies, C36

B

bacilli, C73
bacteria, C20, **C64**
    common shapes of, C73
    ecological roles of, C73
    growing colonies of, C82–84
    harmless/harmful, C99
    helpful, C72
    relative size of, C74
    and viruses, difference between, C92
Beijerinck, Martinus, C74
blood
    cancer of the, C35
    donors, C78–79
    human, C60
        serum, C79
    transfusions, C60, C77–80
    types, C78
breakdown of nutrients, C47
bromthymol blue (BTB), C47
Brown, Robert, C59
bubonic plague, C17–19

C

cancer of the blood, C35
carbon dioxide, C47
carriers of disease, **C12**–16
cell biology, C61
cells, **C32**
    animal, C34, C62–63
    cheek, C45–46
    cork, C32
    division of, C35
    microbial, C64–67
    onion, C43–45
    parts of, C59
    plant, C34

Index

red blood cells, C55, C60, C77, C80
and immune response, C77
size of, C55–57
skin, C34, C58
"typical," C58–59
white blood cells, C60, C77, C80
yeast, C47–50, C71
cell membrane, C51–54, C58
cell theory, **C34**
cellular respiration, **C47**
Centers for Disease Control and Prevention (CDC), C108, 112
Chain, Ernst, C98
cheek cells, C45–46
chicken pox, C3, C4, C10
childbed fever (puerperal infection), C35–C36
cholera, C37, C73
classification
of disease-causing organisms and viruses, C75
of microbes, C68–69
of organisms, C71
compound microscopes, C22
contamination, C85
cork, C32
coverslip, C29
culturing microbes, C82–84
cytoplasm, **C51**, C58, C59

D

decomposers, C73
Dietrich, Richard, C105
diphtheria, C12, C90
diplococci, C73
diseases, **C8**
carriers of, C12, C13–C16
caused by microorganisms, C75, C92, C97–98
emerging, **C107**–112
hereditary, C35
infectious, C4–6
vaccinating against, C90–96
division of cells, C35
DNA, C92
drawing microbes, C28, C32

E

Ebola fever, C107
ecological roles of bacteria, C73
egg, de-shelling, C54
electron microscopes, C61
emerging diseases, **C107**–112
epidemiologists, **C12**, C107
Euglena, C27
eyepieces of microscopes, C24

F

five-kingdom classification scheme, C71, C75
Fleming, Alexander, C98
Florey, Howard, C98
flu shots, C91, C95
focusing microscopes, C24, C25
food poisoning, C85
food spoilage, C36
full course of antibiotics, **C99**–102
fungi, C71, C75

G

genetic material, C92
germs, C38, C70. *See also* microbes
germ theory of disease, **C37**, C32–41
green algae, C27, C28, C33

Index

guidelines for doctors and food industry workers, C89

H

Halstead, William Stewart, C39
handkerchiefs, disposable, C39
hand washing, C36, C85–89
Hansen, G. A., C20
Hansen's disease (leprosy), C19–21
hereditary diseases, C35
hints/tips
    for drawing microbes, C28
    staining slides, C80
    using the microscope, C25, C26, C30
    viewing slides, C44
HIV, C60
Hooke, Robert, C32, C41
human blood, C60

I

immune system, **C77**–81, C82
    and vaccination, C92–93
infections, C39, C73, C85
infectious diseases, **C4**–7, C12–16
    emerging/new, **C107**–112
    germ theory of, **C37**
    microbes that cause, C73–75
    reducing risk of, C36 C38–39

K

kingdoms, C71, C75
Kitasato, Shibasaburo, C22
Koch, Robert, C37, C82
Koch's experiment, C38

L

Leeuwenhoek, Anton van, C33
leprosy (Hansen's disease), C19–21
leukemia, C35
light microscopes, C23–24, C77
Lister, Joseph, C38–39
Louisiana Leper Home, C20
Lyme disease, C17

M

maggots, C39
magnification, C22
malaria, C17, C18
"Maracondo fever," C108–112
methyl cellulose, C29
microbes, **C27,** C34, C40
    classifying, C64, C67, C68–69, C70–76
    culturing, C82–84
    discovery of, C31
    drawing, C28, C32
    preventing growth of, C82–84
    protection against, C77
        relative sizes of, C74
        single-celled, C34, C72
Micrographia (Hooke), C32
microorganism, C70
microscopes
    cleaning, C24
    compound, C22
    correct way to use, C22–26
    electron, C61
    eyepieces of, C23, C24
    focusing, C24, C25
    hints/tips for using, C25, C26, C30
    Hooke's, C32
    Leeuwenhoek's, C33
    light, C23–24, C77

Index

objectives on, C24
oil-immersion, C22
rules for handling, C24
scanning electron, C22, C77
transmission electron, C22
types of, C22
microscopists, C76
molds, C71
mosquitos, C18
multicellular organisms, **C34**, C47, C75
mushrooms, C71
mycobacteria, C73
Mycobacterium leprae, C20
Mycobacterium tuberculosis, C72, C105

N

nematodes, C27
Newton, Isaac, C41
Nightingale, Florence, C38
nucleus, C58, **C59**, C72
nutrients, C47

O

objectives (on microscope), C24
oil-immersion microscopes, C22
onion cells, C43–45
oral vaccines, C93
organelles, **C59**–60
organ transplants, C78
outbreaks of diseases, C13, **C107**–112
overprescribing/improper use of antibiotics, C103–106
oxygen, C47

P

Paramecium, C27, C72
Pasteur, Louis, C36–37, C40
pasteurization, C36
petri dishes, **C82**
plant cells, C34
Plasmodium, C18
pneumonia, C73, C94
polio vaccines, C92–93, C96
protists, **C64**, C71, C72, C74, C75, C76
relative size of, C74
public service announcements (PSAs), C8, C10
puerperal infection (childbed fever), C35–36

Q

quarantine, C19

R

red blood cells, C55, C60, C77, C80
Redi, Francesco, C39
reproduction of viruses, C74
resistant bacteria, **C99**
RNA, C92
Roosevelt, Franklin D., C92–93
rubber gloves, C39, C89
rules for handling a microscope, C24

S

scanning electron microscopes, C22, C77

Schleiden, Matthais Jakob, C34, C42
Schwann, Theodor, C34, C42
Semmelweiss, Ignaz Philipp, C35–36, C85
serum, C79
Siebold, Karl Theodor Ernst von, C34, C42
silk production, C37
size
   of cells, C55–57, C72
   of microbes, C74–75
skin cells, C34, C58
sleeping sickness, C72
Spallanzani, Lazzaro, C40
spirilla, C73
spirochetes, C73
Spirogyra, C28, C33
spoiled food/milk, C36
spontaneous generation, theory of, C39–40
staining slides, C44–45, C80
standard for comparison, C49
Staphylococci, C73
Stentor, C27
sterilization, C39
strep throat, C73
streptococci, C36, C73
Streptococcus pneumoniae, C105
survey questions, C106
symptoms of illnesses, C13, C93
syphilis, C73

T

tetanus, C92
throat cultures, C94
ticks, C17

tips. *See* hints/tips
tobacco plants, C74
transmission electron microscopes, C22
Trypanosoma, C72
tuberculosis (TB), C37, C73, C105
Turlington, Christy, C8
typhoid, C12

V

vaccines/vaccinations, C12, **C90**–96
vectors, **C17,** C18, C20
vibrio, C73
viral diseases, C95
Virchow, Rudolf Carl, C34–35
viruses, **C74**–75, C76, C94
   and bacteria, difference between, C92
   relative size of, C74
   reproduction of, C74
   HIV, C60
   vaccines for, C92

W

washing hands, C36, C85–89
white blood cells, C60, C77, C80

Y

yeast cells, C47–50, C71
Yersin, Alexandre, C22
yogurt, C72

Credits

Abbreviations: t (top), m (middle), b (bottom), l (left), r (right)

All illustrations by Seventeenth Street Studios.

"Talking It Over" icon photo: ©Michael Keller/The Stock Market

Unit opener (C-2, C-3): tm: ©Charles O'Rear/CORBIS; lb: ©Science Pictures Limited/CORBIS; mb: ©Premium Stock/CORBIS; m: ©Ed Eckstein/CORBIS; rb: ©Jack Fields/CORBIS; rm: Dennis Kunkel/PHOTOTAKE

C-4 ©2001 Stuart McClymont/Stone; C-8 Image courtesy from the Centers for Disease Control with permission from Christy Turlington; C-10 Michael Brill, Louisville, Kentucky; C-12 Will & Deni McIntyre/Photo Researchers, Inc.; C-18 ©Richard T. Nowitz/PHOTOTAKE; C-19 Image courtesy of TDR image library, TDR Communications and the World Health Organization; C-20 Philip Gould/CORBIS; C-22 l: ©2001 Stewart Cohen/Stone, r: Dr. Dennis Kunkel/PHOTOTAKE; C-27 tl: ©M.I. Walker/Photo Researchers, Inc., tm, tr, bl, bm: ©Barry Runk/Stan/Grant Heilman Photography, br: ©Eric Grave/PHOTOTAKE; C-28 ©Barry Runk/Stan/Grant Heilman Photography; C-32 ml: ©Bettmann/CORBIS, bl: ©Charles O'Rear/CORBIS, br: ©Lester V. Bergman/CORBIS; C-33 tl: ©Bettmann/CORBIS, mr: ©Science Pictures Limited/CORBIS; C-34 tl: Sue Boudreau, ml: ©Science Pictures Limited/CORBIS, bl: Dr. Dennis Kunkel/PHOTOTAKE; C-35 "Leo the cat": Sylvia Parisotto; C-36 Dr. Dennis Kunkel/PHOTOTAKE; C-38 ml: ©Bettman/CORBIS, bl: ©Bettmann/CORBIS; C-39 ©G. Watson/Photo Reseachers, Inc.; C-42 ©Bettmann/CORBIS; C-47 ©2001 Tracy Frankel/The Image Bank; C-50 ©Barry Runk/Stan/Grant Heilman Photography; C-51 ©Jim Zuckerman/CORBIS; C-58 LesterV. Bergman/CORBIS; C-60 ©Dr. Dennis Kunkel/PHOTOTAKE; C-62 tl: ©CNRI/PHOTOTAKE, tr: ©CNRI/Photo Researchers, Inc., bl: ©Lester V. Bergman/CORBIS, br: ©Tina Carvalho; C-64 tl, tr, bl: ©Eric Grave/PHOTOTAKE, mr: ©Carolina Biological Supply Co./PHOTOTAKE; C-77 r: ©Carolina Biological Supply Co./PHOTOTAKE, l: ©NCI/Photo Researchers, Inc.; C-80 ©Dr. Dennis Kunkel/PHOTOTAKE; C-82 ©Bruce Iverson/DoctorStock.com; C-89 "Guidelines for Doctors Prior to Surgery" source: Thomas Barber, MD, "Guidelines for Food Industry Workers" source: Utah Department of Health; C-90 ©2001 Richard Price/FPG; C-93 ©CORBIS; C-97 ©C. James Webb/PHOTOTAKE; C-107 ©Will & Deni McIntyre/Photo Researchers, Inc.

Cover photo (kids running): ©2001 David Young-Wolff/Stone